ƒP

Also by Martin A. Nowak

Evolutionary Dynamics:
Exploring the Equations of Life (Winner of R. R. Hawkins Award)

WITH ROBERT MAY

Virus Dynamics:
Mathematical Principles of Immunology and Virology

Also by Roger Highfield

The Physics of Christmas:
From the Aerodynamics of Reindeer to the Thermodynamics of Turkey

The Science of Harry Potter:
How Magic Really Works

WITH PETER COVENEY

The Arrow of Time:
A Voyage Through Science to Solve Time's Greatest Mystery
(*New York Times* Notable Book, UK bestseller)

Frontiers of Complexity:
The Search for Order in a Chaotic World

WITH PAUL CARTER

The Private Lives of Albert Einstein

WITH SIR IAN WILMUT

After Dolly:
The Uses and Misuses of Human Cloning

SuperCooperators

*Altruism, Evolution, and Why We Need
Each Other to Succeed*

Martin A. Nowak

with Roger Highfield

FREE PRESS
New York   London   Toronto   Sydney

FREE PRESS
A Division of Simon & Schuster, Inc.
1230 Avenue of the Americas
New York, NY 10020

First Free Press hardcover edition March 2011

FREE PRESS and colophon are trademarks of Simon & Schuster, Inc.

For information about special discounts for bulk purchases,
please contact Simon & Schuster Special Sales at 1-866-506-1949
or business@simonandschuster.com.

The Simon & Schuster Speakers Bureau can bring authors to your live event.
For more information or to book an event contact the Simon & Schuster Speakers Bureau
at 1-866-248-3049 or visit our website at www.simonspeakers.com.

DESIGNED BY ERICH HOBBING

Manufactured in the United States of America

1   3   5   7   9   10   8   6   4   2

Library of Congress Cataloging-in-Publication Data

Nowak, M. A. (Martin A.)
SuperCooperators : altruism, evolution, and why we need each other to succeed /
Martin A. Nowak, Roger Highfield.
p. cm.
Includes bibliographical references.
1. Game theory. 2. Evolution (Biology)—Mathematical models.
3. Cooperative societies. I. Highfield, Roger. II. Title.
QA269.N687 2011
519.3—dc22   2010035517

ISBN 978-1-4391-0018-9
ISBN 978-1-4391-1017-1 (ebook)

To Karl and Bob,
relentless cooperators

CONTENTS

Contents

The only thing that will redeem mankind is cooperation.
—BERTRAND RUSSELL

The Struggle

From the war of nature, from famine and death, the most exalted object which we are capable of conceiving, namely, the production of the higher animals, directly follows.

—Charles Darwin, *On the Origin of Species*

Biology has a dark side. Charles Darwin referred to this shadowy aspect of nature as the struggle for existence. He realized that competition is at the very heart of evolution. The fittest win this endless "struggle for life most severe" and all others perish. In consequence, every creature that crawls, swims, and flies today has ancestors that once successfully thrived and reproduced more often than their unfortunate competitors. As for the rest, they forfeited any chance to contribute to the next generation. They lost, and now they're gone.

The struggle was born at least 4 billion years ago, with the first primitive cells. They were simple bacteria, each one little more than a tiny, organized collection of chemicals. If one of these chemical machines had an advantage over its peers, it would reproduce faster. Given better-than-average access to a limited food source, it would prosper and its rivals perish. This struggle continues, and across a spectrum of habitats. Today, Earth is the planet of the cell. Microorganisms now teem in almost every habitat, from poles to deserts to geysers, rocks, and the inky depths of the oceans. Even in our own bodies, bacterial cells outnumber our own. When adding up the total number of cells on Earth today—around 10 to the power of 30, or 1 followed by 30 zeroes—all you have to do is estimate the number of bacterial cells; the rest is pocket change.

The struggle can also be found in those organized collections of cells that we call animals. On the African savannah, a lion crouches in the long grass, muscles tensed and senses tightly focused on a nearby herd. Slowly and silently it stalks the antelope and then suddenly, in a burst of speed, sprints toward an animal, leaps, grabs its neck, and pierces the skin, blood vessels, and windpipe with its long, sharp teeth. It drags the prey to the ground and holds tight until the antelope breathes its last. When the lion finishes with its kill, a shroud of vultures wraps the bloody remains.

In *The Descent of Man,* Darwin remarked that modern man was born of the same struggle on the same continent. "Africa was formerly inhabited by extinct apes closely allied to the gorilla and chimpanzee; and as these two species are now man's nearest allies, it is more probable that our early progenitors lived on the African continent than elsewhere." Our ancestors spread out to colonize the Earth during the last 60,000 years or so, outcompeting archaic species such as *Homo erectus* and the big-brained Neanderthals (though if you are European, Asian, or New Guinean, you may have a trace of Neanderthal blood racing through your veins). The struggle for existence continues apace, from competition between supermarkets to drive down prices to cutthroat rivalry between Wall Street firms.

In the game of life we are all driven by the struggle to succeed. We all want to be winners. There is the honest way to achieve this objective. Run faster than the pack. Jump higher. See farther. Think harder. Do better. But, as ever, there is the dark side, the calculating logic of self-interest that dictates that one should never help a competitor. In fact, why not go further and make life harder for your rivals? Why not cheat and deceive them too? There's the baker who palms you off with a stale loaf, rather than the one fresh out of the oven. There's the waiter who asks for a tip when the restaurant has already added a service charge. There's the pharmacist who recommends a well-known brand, when you can get a generic version of the same drug much more cheaply. Nice guys finish last, after all.

Humans are the selfish apes. We're the creatures who shun the needs of others. We're egocentrics, mercenaries, and narcissists. We look after number one. We are motivated by self-interest alone, down to every last bone in our bodies. Even our genes are said to be selfish. Yet competition does not tell the whole story of biology. Something profound is missing.

Creatures of every persuasion and level of complexity cooperate to live. Some of the earliest bacteria formed strings, where certain cells in each living filament die to nourish their neighbors with nitrogen. Some bacteria hunt in groups, much as a pride of lions hunt together to corner an antelope; ants form societies of millions of individuals that can solve complex problems, from farming to architecture to navigation; bees tirelessly harvest pollen for the good of the hive; mole rats generously allow their peers to dine on their droppings, providing a delicious second chance to digest fibrous roots; and meerkats risk their lives to guard a communal nest.

Human society fizzes with cooperation. Even the simplest things that we do involve more cooperation than you might think. Consider, for example, stopping at a coffee shop one morning to have a cappuccino and croissant for breakfast. To enjoy that simple pleasure could draw on the labors of a small army of people from at least half a dozen countries.

Farmers in Colombia grew the beans. Brazil provided the lush green fields of swaying sugar cane that was used to sweeten the beverage. The dash of creamy milk came from cows on a local farm and was heated with the help of electricity generated by a nuclear power station in a neighboring state. The barista, being a pretentious sort of fellow, made the coffee with mineral water from Fiji. As for that flaky croissant, the flour came from Canada, the butter from France, and the eggs from a local cooperative. The pastry was heated and browned in a Chinese-made oven. Many more people worked in supply lines that straddle the planet to bring these staples together.

Delivering that hot coffee and croissant also relied on a vast number of ideas, which have been widely disseminated by the remarkable medium of language. The result is a tightly woven network of cooperation stretching across the generations, as great ideas are generated, passed on, used, and embellished, from the first person to drink a beverage based on roasted seeds to the invention of the light bulb that illuminates the coffee shop, to the patenting of the first espresso machine.

The result, that simple everyday breakfast, is an astonishing cooperative feat that straddles both space and time. That little meal relies on concepts and ideas and inventions that have been passed down and around among vast numbers of people over hundreds, even thousands of years. The modern world is an extraordinary collective enterprise. The knowledge of how to select beans, make flour, build ovens, and froth milk is splintered in hundreds of heads. Today, the extent to which our brains collaborate matters as much as the size of our brains.

This is the bright side of biology. The range and the extent to which we work together make us supreme cooperators, the greatest in the known universe. In this respect, our close relatives don't even come close. Take four hundred chimpanzees and put them in economy class on a seven-hour flight. They would, in all likelihood, stumble off the plane at their destination with bitten ears, missing fur, and bleeding limbs. Yet millions of us tolerate being crammed together this way so we can roam about the planet.

Our breathtaking ability to cooperate is one of the main reasons we have managed to survive in every ecosystem on Earth, from scorched, sun-baked deserts to the frozen wastes of Antarctica to the dark, crushing ocean depths. Our remarkable ability to join forces has enabled us to take the first steps in a grand venture to leave the confines of our atmosphere and voyage toward the moon and the stars beyond.

By *cooperation,* I mean more than simply working toward a common aim. I mean something more specific, that would-be competitors decide to aid each other instead. This does not seem to make sense when viewed from a traditional Darwinian perspective. By helping another, a competitor hurts its own fitness—its rate of reproduction—or simply blunts its competitive edge. Yet it is easy to think of examples: a friend drives you to the dentist though it makes her late for work; you donate fifty dollars to charity rather than spending it on yourself. The cells in your body, rather than reproduce willy nilly to selfishly expand their own numbers, respect the greater needs of the body and multiply in an orderly fashion to create the kidney, the liver, the heart, and other vital organs.

Many everyday situations can be viewed as choices about whether or not to cooperate. Let's say you want to open a savings account with a British bank (as we discovered in *Mary Poppins,* which appeared long before the credit crunch, "a British bank is run with precision"). Imagine that you are standing at the counter as a smiling clerk patiently explains the various options on offer. Banks like to confuse their customers by offering a large number of accounts that differ in terms of fees, interest rates, access, and conditions. If you ask for the best interest rate, the clerk can interpret this apparently simple question in two ways. From his point of view, the best interest rate is the most meager and restrictive, the one that earns the bank the maximum profit. From the customer's point of view, the best rate is the one that earns the most money. If the clerk offers the former, that is an example of defection. But if he recommends an account that gives you, and not the bank, the maximum return, that is an example of cooperation.

Once cooperation is expressed in this way, it seems amazing. Why weaken your own fitness to increase the fitness of a competitor? Why bother to look after anyone besides number one? Cooperation goes against the grain of self-interest. Cooperation is irrational. From the perspective of Darwin's formulation for the struggle for existence, it makes no sense to aid a potential rival, yet there is evidence that this occurs among even the lowliest creatures. When one bacterium goes to the trouble of making an enzyme to digest its food, it is helping to feed neighboring cells too—rivals in the struggle to survive.

This looks like a fatal anomaly in the great scheme of life. Natural selection should lead animals to behave in ways that increase their own chances of survival and reproduction, not improve the fortunes of others. In the never-ending scrabble for food, territory, and mates in evolution, why would one individual ever bother to go out of its way to help another?

BEYOND COOPERATION

We are all dependent on one another, every soul of us on earth.
—George Bernard Shaw, *Pygmalion*

Scientists from a wide range of disciplines have attempted for more than a century to explain how cooperation, altruism, and self-sacrifice arose in our dog-eat-dog world. Darwin himself was troubled by self-less behavior. Yet in his great works, the problem of cooperation was a sideshow, a detail that had to be explained away. That attitude prevails among many biologists even today.

In stark contrast, I believe that our ability to cooperate goes hand in hand with succeeding in the struggle to survive, as surmised more than a century ago by Peter Kropotkin (1842–1921), the Russian prince and anarchist communist who believed that a society freed from the shackles of government would thrive on communal enterprise. In *Mutual Aid* (1902), Kropotkin wrote: "Besides *the law of Mutual Struggle* there is in Nature *the law of Mutual Aid,* which, for the success of the struggle for life, and especially for the progressive evolution of the species, is far more important than the law of mutual contest. This sugges-tion . . . was, in reality, nothing but a further development of the ideas expressed by Darwin himself."

I have spent more than two decades cooperating with many great minds to solve the mystery of how natural selection can lead to mutual aid, so that competition turns into cooperation. I have introduced some new ideas to this well-explored field and refined this mix with my own specialty, which relies on blending mathematics and biology. My studies show that cooperation is entirely compatible with the hard-boiled arithmetic of survival in an unremittingly cold-eyed and com-petitive environment. Based on mathematical insights, I have created idealized communities in a computer and charted the conditions in which cooperation can take hold and bloom. My confidence in what I have found has been bolstered by research on a wide range of species, from bugs to people. In light of all this work, I have now pinned down five basic mechanisms of cooperation. The way that we human beings collaborate is as clearly described by mathematics as the descent of the apple that once fell in Newton's garden.

These mechanisms tell us much about the way the world works. They reveal, for example, that your big brain evolved to cope with gossip, not the other way around; that your guts have cone-like glands to fend off that potentially deadly breakdown of cellular cooperation that we know as cancer; that you are more generous if you sense that you are being watched (even if you are not); that the fewer friends you have, the more strongly your fate is bound to theirs; genes may not be that selfish, after all; if you are a cooperator, you will find yourself surrounded by other cooperators so that what you reap is what you sow; no matter what we do, empires will always decline and fall; and to succeed in life, you need to work together—pursuing the snuggle for existence, if you like—just as much as you strive to win the struggle for existence. In this way, the quest to understand cooperation has enabled us to capture the essence of all kinds of living, breathing, red-blooded evolving processes.

Cooperation—not competition—underpins innovation. To spur creativity, and to encourage people to come up with original ideas, you need to use the lure of the carrot, not fear of the stick. Cooperation is the architect of creativity throughout evolution, from cells to multicellular creatures to anthills to villages to cities. Without cooperation there can be neither construction nor complexity in evolution.

I can derive everyday insights—as well as many unexpected ones—from mathematical and evolutionary models of cooperation. While the idea that the trajectory of spears, cannonballs, and planets can be traced out by equations is familiar, I find it extraordinary that we can also use mathematics to map out the trajectory of evolution. And, of course, it is one thing to know how to foster cooperation but it is quite another to explain why an action helps us get along with each other and to what extent. The mathematical exploration of these mechanisms enables us to do this with profound understanding and with precision too. This is proof, as if we need it, that math is universal.

In the following chapters I will explain the origins of each mechanism of cooperation and interweave this train of thought with my own intellectual journey, one that began in Vienna and then continued to Oxford, Princeton, and now Harvard. En route, I have had the honor to cooperate with many brilliant scientists and mathematicians. Two of them proved particularly inspirational: Karl Sigmund and Robert May, for reasons that will become clear. I have also had to enlist the help of computer programs, students willing to play games, and various funding bodies, from foundations to philanthropists. It is a lovely and intoxicating thought that a high degree of cooperation is required to understand cooperation. And to further underline this powerful idea, this book is also a feat of cooperation between Roger Highfield and myself.

The implications of this new understanding of cooperation are profound. Previously, there were only two basic principles of evolution—mutation and selection—where the former generates genetic diversity and the latter picks the individuals that are best suited to a given environment. For us to understand the creative aspects of evolution, we must now accept that cooperation is the third principle. For selection you need mutation and, in the same way, for cooperation you need both selection and mutation. From cooperation can emerge the constructive side of evolution, from genes to organisms to language and complex social behaviors. Cooperation is the master architect of evolution.

My work has also shown that cooperation always waxes and wanes. The degree to which individuals are able to cooperate rises and falls, like the great heartbeat of nature. That is why, even though we are extraordinary cooperators, human society has been—and always will be—riven with conflict. Global human cooperation now teeters on a threshold. The accelerating wealth and industry of Earth's increasing inhabitants—itself a triumph of cooperation—is exhausting the ability of our home planet to support us all. There's rising pressure on each of us to compete for the planet's dwindling resources.

Many problems that challenge us today can be traced back to a profound tension between what is good and desirable for society as a whole and what is good and desirable for an individual. That conflict can be found in global problems such as climate change, pollution, resource depletion, poverty, hunger, and overpopulation. The biggest issues of all—saving the planet and maximizing the collective lifetime of the species *Homo sapiens*—cannot be solved by technology alone. They require novel ways for us to work in harmony. If we are to continue to thrive, we have but one option. We now have to manage the planet as a whole. If we are to win the struggle for existence, and avoid a precipitous fall, there's no choice but to harness this extraordinary creative force. We now have to refine and to extend our ability to cooperate. We must become familiar with the science of cooperation. Now, more than ever, the world needs SuperCooperators.

SuperCooperators

The Prisoner's Dilemma

I believe that mathematical reality lies outside us, that our function is to discover or observe it, and that the theorems which we prove, and which we describe grandiloquently as our "creations," are simply the notes of our observations.

—Godfrey H. Hardy,
A Mathematician's Apology

At first, I did not appreciate the point of mathematics. I played with numbers during lessons in high school. I enjoyed solving problems. Arithmetic lessons were fun. Math was, all in all, quite interesting. But it was unclear to me what it was for. Perhaps it was a kind of mental gymnastics that had been devised—along with Latin—with the express purpose of making the children's lives just that little bit harder.

At university I changed my mind. I had an epiphany, a spine-tingling moment when I realized that the precisely defined terms, equations, and symbols of mathematics are fundamental. I came to realize that mathematics holds the key to formulating the laws that govern the cosmos, from the grandest filaments, voids, and structures that stretch across the heavens to the peculiar behavior of the tiniest and most ubiquitous grains of matter. More important, it could say something profound about everyday life.

Mathematics is characterized by order and internal consistency as well as by numbers, shapes, and abstract relationships. Although you might feel that these concepts only inhabit the human mind, some of them are so real and absolute that they mean precisely the same thing to us as they would to a clever many-tentacled alien floating on an icy exoplanet on the far side of the universe. In fact, I would go even further than saying the ideas of mathematics are objective and concrete. The cosmos itself is mathematical: everything and anything that happens in it is the consequence of universal logic acting on universal rules.

Beyond the dimensions of space and time, mathematics inhabits a nonmaterial realm, one that is eternal, unchanging, and ever true. The empire of mathematics extends far beyond what we can see around us, beyond what we are able to perceive, and far beyond what we can imagine. There's an unseen, perfect, and transcendental universe of possibilities out there. Even in the wake of cosmic degradation, collapse, and ruin, the inhabitants of other universes will still be there to gaze on the unending beauty of mathematics, the very syntax of nature. The truth really is out there and it can be expressed in this extraordinary language.

Some would go even further than this. They regard the mathematics that describes our cosmos as a manifestation of the thoughts of a creator. Albert Einstein once remarked: "I believe in Spinoza's God, Who reveals Himself in the lawful harmony of the world." For the seventeenth-century Dutch philosopher who had so impressed Einstein, God and nature were as one (*deus sive natura*), and the practice of doing math was tantamount to a quest for the divine. Whenever I think about this connection, I am always reminded of the last, thrilling lines of Goethe's *Faust*:

> *All that is changeable / Is but refraction*
> *The unattainable / Here becomes action*
> *Human discernment / Here is passed by*
> *Woman Eternal / Draws us on high.*

My epiphany at university was that somewhere in this infinite, unimaginable ocean of truth there is a corporeal mathematics, a splash of math that you can feel, smell, and touch. This is the mathematics of the tangible, from the equations that govern the pretty patterns formed by the red petals of a rose to the laws that rule the sweeping movements of Mars, Venus, and other planets in the heavens. And of all those remarkable insights that it offers, I discovered that mathematics can capture the quintessence of everyday life, the ever-present tension that exists between conflict and cooperation.

This tension is palpable. It tugs at the emotions of participants in an internet purchase, where there is a temptation for buyers not to pay for goods and sellers not to send them. The tension surfaces when weighing whether to contribute to the public good, whether through taxes or licenses, or whether to clear up after a picnic on the beach or sort out items of everyday rubbish that can be recycled. One can feel this strain between the personal and public in transport systems too, which trust that enough people will pay for a ticket to ensure that they can operate sufficient buses, trains, and trams.

This tension between the selfish and selfless can be captured by the Prisoner's Dilemma. Although it is a simple mathematical idea, it turns out to be an enchanted trap that has ensnared some of the brightest minds for decades. I myself became so infatuated with playing this extraordinary mathematical game that I changed my course at university and, at a stroke, changed the course of my life.

My work on the Dilemma gave me the first critical insights into why our traditional understanding of evolution is incomplete. It revealed why, in addition to the fundamental forces of mutation and selection, we need a third evolutionary force, that of cooperation. It provided a way to hone my understanding of the mechanisms that make someone go out of her way to help another. The Dilemma has played a key role in cementing the foundations for an understanding of the future of human cooperation.

PRISONER OF THE DILEMMA

As a schoolboy, I wanted to be a doctor. Then I read *The Eighth Day of Creation: Makers of the Revolution in Biology* (1979) by *Time* magazine journalist Horace Judson. This wonderful chronicle of the birth of molecular biology put an end to my medical ambitions. I made up my mind there and then to study the very chemical basis of life, the molecules that build our cells, power them, organize them, and run them. I would pursue biochemistry at the University of Vienna. Not everyone was enthusiastic about my decision. My parents were troubled by my move away from a career as a medical doctor, a guaranteed way to become a respected pillar of society. Their only child was now going to study a subject that, as far as they were concerned, had mostly to do with yeast, which was central to fermenting beer and wine.

In October 1983 I walked into my first lecture and encountered "girls"—many more than I had ever seen before and conveniently all in one place. Thanks to the female-dominated intake of a pharmacology course, girls made up nearly two-thirds of the six hundred people now crammed around me in the lecture hall. Having been educated at an all-boys school, I thought I was in paradise. Among the handful of chemistry students was Ursula, who like me was struggling to keep pace with the university's intensive introduction to mathematics. Six years later, we were married. I still wonder whether I was selected for my ability to solve mathematical problems.

As I became besotted at the University of Vienna, the emphasis of my studies gradually changed. I adored physics in the first year, then physical chemistry in the second year. In the third year, I had the great good fortune to be lectured on theoretical chemistry by the formidable Peter Schuster, who helped to establish the Viennese school of mathematical biology and, later, would become the president of the illustrious Austrian Academy of Sciences and deliver a lecture to Pope Benedict XVI on the science of evolution. I knew immediately that I wanted to work with Peter. In the fourth year, I began to study with him for a diploma thesis. An ebullient character, he was supremely knowledgeable and his interests extended well beyond science. Once, when we went mountain climbing together, he declared: "There's no such thing as bad weather, only insufficient equipment."

The moment when I realized that I was well and truly smitten by mathematics came a year later, while on an Alpine jaunt with Peter. It was March 1988, during my early days as a doctoral student, and I was on a retreat. With me was a fresh crop of talent, including Walter Fontana, who today is a prominent biologist at Harvard Medical School. Our group was staying in a primitive wooden hut in the Austrian mountains to enjoy lots of fresh air, work, and play. We skied, we listened to lectures, we drank beer and wine, and we contemplated the mysteries of life. Best of all, we discussed new problems and theory, whether in the cozy warmth of the little hut or outside, in the chilled Alpine air. As the ideas tumbled out at high altitude, our breath condensed into vapor. I can't remember if they were mathematical dreams or just clouds of hot air. But the experience was exhilarating.

The mix of bright-eyed students was enriched with impressive academics. Among them was Karl Sigmund, a mathematician from the University of Vienna. With his wild shock of hair, bottle-brush mustache, and spectacles, Karl looked aloof and unapproachable. He was cool, more like a student than a professor. Karl would deliver all his lectures from memory with a hypnotic, almost incantatory rhythm. On the last day of that heady Alpine meeting, he gave a talk on a fascinating problem that he himself had only just read about in a newspaper article.

The article described work in a field known as game theory. Despite some earlier glimmerings, most historians give the credit for developing and popularizing this field to the great Hungarian-born mathematician John von Neumann, who published his first paper on the subject in 1928. Von Neumann went on to hone his ideas and apply them to economics with the help of Oskar Morgenstern, an Austrian economist who had fled Nazi persecution to work in the United States. Von Neumann would use his methods to model the cold war interaction between the United States and the Soviet Union. Others seized on this approach too, notably the RAND Corporation, for which von Neumann had been a consultant. The original "think tank," the RAND (*R*esearch *an*d *D*evelopment) Corporation was founded as Project RAND in December 1945 by the U.S. Army Air Force and by defense contractors to think the unthinkable.

In his talk, Karl described the latest work that had been done on the Prisoner's Dilemma, an intriguing game that was first devised in 1950 by Merrill Flood and Melvin Dresher, who worked at RAND in Santa Monica, California. Karl was excited about the Dilemma because, as its inventors had come to realize, it is a powerful mathematical cartoon of a struggle that is central to life, one between conflict and cooperation, between the individual and the collective good.

The Dilemma is so named because, in its classic form, it considers the following scenario. Imagine that you and your accomplice are both held prisoner, having been captured by the police and charged with a serious crime. The prosecutor interrogates you separately and offers each of you a deal. This offer lies at the heart of the Dilemma and goes as follows: If one of you, the defector, incriminates the other, while the partner remains silent, then the defector will be convicted of a lesser crime and his sentence cut to one year for providing enough information to jail his partner. Meanwhile, his silent confederate will be convicted of a more serious crime and burdened with a four-year sentence.

If you both remain silent, and thus cooperate with each other, there will be insufficient evidence to convict either of you of the more serious crime, and you will each receive a sentence of two years for a lesser offense. If, on the other hand, you both defect by incriminating each other, you will both be convicted of the more serious crime, but given reduced sentences of three years for at least being willing to provide information.

In the literature, you will find endless variants of the Dilemma in terms of the circumstances, the punishments and temptations, the details of imprisonment, and so on. Whatever the formulation, there is a simple central idea that can be represented by a table of options, known as a payoff matrix. This can sum up all four possible outcomes of the game, written down as two entries on each of the two lines of the matrix. This can sum up the basic tensions of everyday life too.

Let's begin with the top line of the payoff matrix: You both cooperate (that means a sentence of two years each and I will write this as −2 to underline the years of normal life that you lose). You cooperate and your partner defects (−4 years for you, −1 for him). On the second line come the other possible variants: You defect, and your partner cooperates (−1 for you, −4 for him). You both defect (−3 years each). From a purely selfish point of view, the best outcome for you is the third, then the first, then the fourth, and finally the second option. For your confederate the second is the best option, followed by the first, fourth, and third.

Payoff Matrix			
		opponent	
		cooperate	defect
player	cooperate	-2,-2	-4,-1
	defect	-1,-4	-3,-3

What should you do, if you cast yourself as a rational, selfish individual who looks after number one? Your reasoning should go like this. Your partner will either defect or cooperate. If he defects, you should too, to avoid the worst possible outcome for you. If he cooperates, then you should defect, as you will get the smallest possible sentence, your preferred outcome. Thus, no matter what your partner does, it is best for you to defect.

Defecting is called a dominant strategy in a game with this payoff matrix. By this, the theorists mean that the strategy is always the best one to adopt, regardless of what strategy is used by the other player. This is why: If you both cooperate, you get two years in prison but you only get one year in prison if you defect. If the other person defects and you hold your tongue, then you get four years in prison, but you only get three years if you both defect. Thus no matter what the other person does, it is better for you to defect.

But there's a problem with this chain of reasoning. Your confederate is no chump and is chewing over the Dilemma in precisely the same way as you, reaching exactly the same conclusion. As a consequence, you both defect. That means spending three years in jail. The Dilemma comes because if you both follow the best, most rational dominant strategy it leaves both of you worse off than if you had both remained silent! You both end up with the third best outcome, whereas if you had both cooperated you would have both enjoyed the second best outcome.

That, in a bitter nutshell, is the Prisoner's Dilemma. If only you had trusted each other, by cooperating, you would both be better off than if you had both acted selfishly. With the help of the Dilemma, we can now clearly appreciate what it means to cooperate: one individual pays a cost so that another receives a benefit. In this case, if both cooperate, they forfeit the best outcome—a one-year sentence—and both get second best. This is still a better result than either of you can achieve if you both defect.

To create the Dilemma, it is important to arrange the relative size of each of the payoffs for cooperation and defection in the matrix in the correct way. The Dilemma is defined by the exact ranking of the payoff values, where R is the reward for mutual cooperation; S is the sucker's payoff for cooperating when your fellow player defects; T is the temptation to renege when your fellow player cooperates, and P is the punishment if both players defect. Let's spell this out. When the players both cooperate, the payoff (R) is greater than the punishment (P) if they both defect. But when one cooperates and one defects, the person who is tempted to renege gets the highest payoff (T) while the hapless cooperator ends up with the lowest of all, the sucker's payoff (S). Overall, we can create the Dilemma if T is greater than R which is greater than P which is greater than S. We can rank the payoffs in the basic game in other, different ways and still end up with cooperative dilemmas. But of all of them, the Prisoner's Dilemma is by far the hardest to solve. You can think of it as the ultimate dilemma of cooperation.

We all encounter the Dilemma in one form or another all the time in everyday life. Do I want to help a competitor in the office—for instance, offer to do his work during his holiday—when this person is competing with me for a promotion? When two rival firms set prices, should they both go for as much as they can, colluding in some way, or should one company try to undercut its competitor? Arms races between superpowers, local rival nations, or even different species offer other examples of the Dilemma at work. Rival countries are better off when they cooperate to avoid an arms race. Yet the dominant strategy for each nation is to arm itself heavily. And so on and so forth.

INCARCERATION

On my first encounter with the Prisoner's Dilemma in that Alpine hut, I was transfixed. By that time, Karl had actually become my prisoner. He didn't have any transport and I offered him a ride back to Vienna. We discussed the Dilemma as we drove back the next day in the same VW that my father still uses today to putter around Austria. Even after I dropped Karl off, I kept him in my sights. Before long, I was doing a PhD with him at the Institute for Mathematics in Vienna. Students who had studied there before me include the great physicist Ludwig Boltzmann, the logician Kurt Gödel, and the father of genetics, Gregor Mendel.

As I pursued my doctorate, Karl and I would often meet in local coffeehouses, the genius loci of past glory. In these inspiring surroundings Gödel had announced his incompleteness theorem, Boltzmann had worked on entropy, and Wittgenstein had challenged the Vienna Circle, a group of intellectuals who would gather to discuss mathematics and philosophy. One day we sat in the Café Central, an imposing building with arched ceilings and marble columns, where Trotsky had planned the Russian revolution.

As we sipped thick, strong coffee and chatted about how to solve the Prisoner's Dilemma, Karl and I rediscovered the subtleties of a problem that had transfixed bright minds for generations. Little did we realize that in the decades that followed, we would devise new mathematics to explore the Dilemma. We would create communities of agents in a computer, study how they evolved, and conduct analyses to reveal the mechanisms able to solve the Dilemma. I would establish teams at Oxford, Princeton, and Harvard as well as collaborations with mathematicians, biologists, chemists, doctors, and economists around the world to understand how these mechanisms worked and what their wider implications were.

Some scientists regard the Prisoner's Dilemma as a remarkably revealing metaphor of biological behavior, evolution, and life. Others regard it as far too simple to take into account all the subtle forces at play in real societies and in biology. I agree with both camps. The Dilemma is not itself the key to understanding life. For the Dilemma to tell us something useful about the biological world, we need to place it in the context of evolution.

Evolution can only take place in populations of reproducing individuals. In these populations, mistakes in reproduction lead to mutation. The resulting mutants might reproduce at different rates, as one mutant does better in one environment than another. And reproduction at different rates leads to selection—the faster-reproducing individuals are selected and thrive. In this context we can think about the payouts of the Prisoner's Dilemma in terms of what evolutionary scientists call "fitness" (think of it as the rate of reproduction). Now we can express what cooperation in the Prisoner's Dilemma means when placed in an evolutionary context: if I help you then I lower my fitness and increase your fitness.

Here's where the story gets fascinating. Now that we have put the Dilemma in an evolutionary form, we discover that there is a fundamental problem. Natural selection actually opposes cooperation in a basic Prisoner's Dilemma. At its heart, natural selection undermines our ability to work together. Why is this? Because in what mathematicians call a well-mixed population, where any two individuals meet equally often, cooperators always have a lower fitness than defectors—they're always less likely to survive. As they die off, natural selection will slowly increase the number of defectors until all the cooperators have been exterminated. This is striking because a population consisting entirely of cooperators has a higher average fitness than a population made entirely of defectors. Natural selection actually destroys what would be best for the entire population. Natural selection undermines the greater good.

To favor cooperation, natural selection needs help in the form of mechanisms for the evolution of cooperation. We know such mechanisms exist because all around us is abundant evidence that it does pay to cooperate, from the towering termite mound to the stadium rock concert to the surge of commuters in and out of a city during a working day. In reality, evolution has used these various mechanisms to overcome the limitations of natural selection. Over the millennia they have shaped genetic evolution, in cells or microbes or animals. Nature smiles on cooperation.

These mechanisms of cooperation shape cultural evolution too, the patterns of change in how we behave, the things we wear, what we say, the art we produce, and so on. This aspect of evolution is more familiar: when we learn from each other and alter the way we act accordingly. It also takes place over much shorter timescales. Think about a population of humans in which people learn different strategies to cope with the world around them, whether religion or boat building or hammering a nail into a piece of wood. The impact of cooperation on culture is huge and, for me, the central reason why life is so beguiling and beautiful.

QUEST FOR THE EVOLUTION
OF COOPERATION

Mathematics, rightly viewed, possesses not only truth, but supreme beauty—a beauty cold and austere, like that of sculpture, without appeal to any part of our weaker nature, without the gorgeous trappings of painting or music, yet sublimely pure, and capable of a stern perfection such as only the greatest art can show.
—Bertrand Russell, *Study of Mathematics*

My overall approach to reveal and understand the mechanisms of cooperation is easy to explain, even if my detailed workings might appear mysterious. I like to take informal ideas, instincts, even impressions of life and render them into a mathematical form. Mathematics allows me to chisel down into messy, complicated issues and—with judgment and a little luck—reveal simplicity and grandeur beneath. At the heart of a successful mathematical model is a law of nature, an expression of truth that is capable of generating awe in the same way as Michelangelo's extraordinary sculptures, whose power to amaze comes from the truth they capture about physical beauty.

Legend has it that when asked how he had created David, his masterpiece, Michelangelo explained that he simply took away everything from the block of marble that was not David. A mathematician, when confronted by the awesome complexity of nature, also has to hack away at a wealth of observations and ideas until the very essence of the problem becomes clear, along with a mathematical idea of unparalleled beauty. Just as Michelangelo wanted his figures to break free from the stone that imprisoned them, so I want mathematical models to take on a life beyond my expectations, and work in circumstances other than those in which they were conceived.

Michelangelo sought inspiration from the human form, notably the male nude, and also from ideas such as Neoplatonism, a philosophy that regards the body as a vessel for a soul that longs to return to God. Over the few centuries that science has been trying to make sense of nature, the inspiration for mathematical representations of the world has changed. At first, the focus was more on understanding the physical world. Think of how Sir Isaac Newton used mathematics to make sense of motion, from the movement of the planets around the sun to the paths of arrows on their way to a target. To the amazement of many, Newton showed that bodies on Earth and in the majestic heavens were governed by one and the same force—gravity—even though planets are gripped in an orbit while objects like arrows and apples drop to the ground.

Today, the models of our cosmos are also concerned with biology and society. Among the eddies and ripples of that great river of ideas that has flowed down the generations to shape the ways in which scientists model these living aspects of the world are the powerful currents generated by Charles Darwin (1809–1882), who devised a unifying view of life's origins, a revolutionary insight that is still sending out shock waves today.

Darwin worked slowly and methodically, using his remarkable ability to make sense of painstaking studies he had conducted over decades, to conclude that all contemporary species have a common ancestry. He showed that the process of natural selection was the major mechanism of change in living things. Because reproduction is not a perfect form of replication, there is variation and with this diversity comes the potential to evolve. But equally, as the game of Chinese Whispers (also known as Gossip or Telephone) illustrates, without a way of selecting changes that are meaningful—a sentence that makes sense—the result is at best misleading and at worst a chaotic babble. Darwin came up with the idea that a trait will persist over many generations only if it confers an evolutionary advantage, and that powerful idea is now a basic tenet of science.

Darwin's message is simple and yet it helps to generate boundless complexity. There exists, within each and every creature, some information that can be passed from one generation to the next. Across a population, there is variation in this information. Because when there are limited resources and more individuals are born than can live or breed, there develops a struggle to stay alive and, just as important, to find a mate. In that struggle to survive, those individuals who bear certain traits (kinds of information) fail and are overtaken by others who are better suited to their environs. Such inherited differences in the ability to pass genes down the generations—natural selection—mean that advantageous forms become more common as the generations succeed. Only one thing counts: survival long enough to reproduce.

Darwin's theory to explain the diverse and ever-changing nature of life has been buttressed by an ever-increasing wealth of data accumulated by biologists. As time goes by, the action of selection in a given environment means that important differences can emerge during the course of evolution. As new variations accumulate, a lineage may become so different that it can no longer exchange genes with others that were once its kin. In this way, a new species is born. Intriguingly, although we now call this mechanism "evolution," the word itself does not appear in *The Origin of Species*.

Darwin himself was convinced that selection was ruled by conflict.

He wrote endlessly about the "struggle for existence" all around us in nature. His theme took on a life of its own as it was taken up and embellished with gusto by many others. Nature is "red in tooth and claw," as Tennyson famously put it when recalling the death of a friend. The catchy term "survival of the fittest" was coined in 1864 by the philosopher Herbert Spencer, a champion of the free market, and this signaled the introduction of Darwinian thinking into the political arena too.

Natural selection is after all about competition, dog-eat-dog and winner takes all. But Darwin was of course talking about the species that was the best adapted to an environment, not necessarily the strongest. Still, one newspaper concluded that Darwin's work showed that "might is right & therefore that Napoleon is right & every cheating tradesman is also right." Darwin's thinking was increasingly abused to justify the likes of racism and genocide, to explain why white colonialists triumphed over "inferior" native races, to breed "superior" humans and so on. These abuses are, in a twisted and depressing way, a testament to the power of his ideas.

But, as I have already stressed, competition is far from being the whole story. We help each other. Sometimes we help strangers too. We do it on a global scale with charities such as Oxfam, which helps people in more than seventy countries, and the Bill & Melinda Gates Foundation, which supports work in more than one hundred nations. We do it elaborately, with expensive celebrity-laden fund-raising dinners in smart venues. We are also charitable to animals. Why? This may look like an evolutionary loose end. In fact it is absolutely central to the story of life.

When cast in an evolutionary form, the Prisoner's Dilemma shows us that competition and hence conflict are always present, just as yin always comes with yang. Darwin and most of those who have followed in his giant footsteps have talked about mutation and selection. But we need a third ingredient, cooperation, to create complex entities, from cells to societies. I have accumulated a wide range of evidence to show that competition can sometimes lead to cooperation. By understanding this, we can explain how cells, and multicellular organisms such as people, evolved, and why they act in the complicated ways that they do in societies. Cooperation is the architect of living complexity.

To appreciate this, we first need to put evolution itself on a firmer foundation. Concepts such as mutation, selection, and fitness only become precise when bolted down in a mathematical form. Darwin himself did not do this, a shortcoming that he was only too aware of. In his autobiography, he confessed his own inability to do sums: "I have deeply regretted that I did not proceed far enough at least to understand something of the great leading principles of mathematics; for men thus endowed seem to have an extra sense." He seemed aware that more rigor was required to flesh out the implications of his radical ideas about life. He regarded his mind "as a machine for grinding general laws out of large collections of facts." But even Darwin yearned for a more "top down" approach, so he could conjure up more precise laws to explain a great mass of data. He needed a mathematical model.

The modern understanding of the process of inheritance is now called "Mendelian," in honor of Gregor Mendel, who had settled for being a monk after failing his botany exams at the University of Vienna. By sorting out the results of crossing round and wrinkly peas, Mendel revealed that inheritance is "particulate" rather than "blending." Offspring inherit individual instructions (genes) from their parents such that round and wrinkly parents produce either round or wrinkly offspring and not something in between. What is often overlooked in his story is that Mendel was a good student of mathematics. The great geneticist and statistician Sir Ronald Fisher went so far as to call him "a mathematician with an interest in biology." Mendel uncovered these rules of inheritance because he was motivated by a clear mathematical hypothesis, even to the extent of ignoring ambiguous results that did not fit. Had Mendel conducted an open-minded statistical analysis of his results, he might not have been successful.

A simple equation to show the effect of passing genes down the generations was found in 1908 by G. H. Hardy, a cricket-loving Cambridge mathematician who celebrated the artistry of his subject in his timeless book *A Mathematician's Apology*. In an unusual reversal of the usual roles, the work of this pure mathematician was generalized by the

German doctor Wilhelm Weinberg to show the incidence of genes in a population. Robert May (now Lord May of Oxford) once went so far as to call the Hardy-Weinberg law biology's equivalent of Newton's first law. Thanks to Hardy and Weinberg we now had a mathematical law that applied across a spectrum of living things.

This attempt to model how inheritance works in nature was extended in seminal investigations conducted in the 1920s and 1930s by a remarkable trio. First, Sir Ronald Fisher, whose extraordinary ability to visualize problems came from having to be tutored in mathematics as a child without the aid of paper and pen, due to his poor eyesight. There was also the mighty figure of J. B. S. Haldane, an aristocrat and Marxist who once edited the *Daily Worker*. I will return to Haldane in chapter 5. The last of this remarkable trio was Sewall Wright, an American geneticist who was fond of philosophy, that relative of mathematics (forgive me for cracking the old joke about the difference: while mathematicians need paper, pencil, and a wastepaper basket, philosophers need only paper and pencil).

Together, this threesome put the fundamental concepts of evolution, selection, and mutation in a mathematical framework for the first time: they blended Darwin's emphasis on individual animals competing to sire the next generation with Mendel's studies of how distinct genetic traits are passed down from parent to offspring, a combination now generally referred to as the synthetic view of evolution, the modern synthesis, or neo-Darwinian. With many others, I have also extended these ideas by looking at the Prisoner's Dilemma in evolving populations to come up with the basic mechanisms that explain how cooperation can thrive in a Darwinian dog-eat-dog world.

Over the years I have explored the Dilemma, using computer models, mathematics, and experiments to reveal how cooperation can evolve and how it is woven into the very fabric of the cosmos. In all there are five mechanisms that lead to cooperation. I will discuss each one of them in the next five chapters and then, in the remainder of the book, show how they offer novel insights into a diverse range of issues, stretching from straightforward feats of molecular cooperation to the many and intricate forms of human cooperation.

I will examine the processes that paved the way to the emergence of the first living things and the extraordinary feats of cooperation that led to multicellular organisms, along with how cellular cooperation can go awry and lead to cancer. I will outline a new theory to account for the tremendous amount of cooperation seen in the advanced social behavior of insects. I will move on to discuss language and how it evolved to be the glue that binds much of human cooperation; the "public goods" game, the biggest challenge to cooperation today; the role of punishment; and then networks, whether of friends or acquaintances, and the extraordinary insights into cooperation that come from studying them. Humans are SuperCooperators. We can draw on all the mechanisms of cooperation that I will discuss in the following pages, thanks in large part to our dazzling powers of language and communication. I also hope to explain why I have come to the conclusion that although human beings are the dominant cooperators on Earth, man has no alternative but to evolve further, with the help of the extraordinary degree of control that we now exert over the modern environment. This next step in our evolution is necessary because we face serious global issues, many of which boil down to a fundamental question of survival. We are now so powerful that we could destroy ourselves. We need to harness the creative power of cooperation in novel ways.

Five Ways to Solve
the Dilemma

CHAPTER 1

Direct Reciprocity— Tit for Tat

It will have blood; they say, blood will have blood.
—Shakespeare, *Macbeth*

In the pitch darkness, the creatures take flight. They shun the moonlight, making the most of their sense of smell to track their victims, then land nearby to stalk them. After a quick loping run on all fours they latch on to their prey. Using a heat sensor on the nose, each one can tell where the blood courses closest to the surface of the victim's skin. Often a meal begins with a quick bite to the neck. There they can hang for up to half an hour, using their long grooved tongues like straws to lap fresh warm blood. Over several nights they return to sup on the same wounds, and it is thought that they are able to recognize the breathing sounds of their victims in the same way as we use the sound of a voice to recognize each other.

What I find most extraordinary of all about vampire bats is what happens when they return to their roost, where hundreds, even thousands of them congregate, suspended upside down. If one member in the roost is unable to find prey during the night's hunt, its peers will regurgitate some of their bloody fare and share it. The exchange of blood among the bats was first revealed in studies conducted in the early 1980s by Gerald Wilkinson of the University of Maryland. During fieldwork in Costa Rica, Wilkinson found that, on any given night, a few percent of adult bats and one-third of juveniles fail to dine. They rarely starve, however, since well-fed vampire bats disgorge a little precious blood to nourish their hungry peers. What was neat was that his experiments suggested that bats are more likely to share blood with a bat that has previously fed them (the bats spend time grooming each other, paying particular attention to fur around the stomach, enabling them to keep tally).

This is an example of what I call direct reciprocity. By this, I mean simply the principle of give-and-take. When I scratch your back, I expect you to scratch mine in return. The same goes for blood meals among bats. This form of reciprocity is recognized in popular sayings, such as "tit for tat" and the idiom "one good turn deserves another." The Romans used the phrase *quid pro quo*—"something for something." As the vampires suggest, this kind of cooperation dates back long before Romulus and Remus, long before the rise of modern humans.

For direct reciprocity to work, both sides have to be repeatedly in contact so that there is an opportunity to repay one act of kindness with another. They might live in the same road, or village. Perhaps they work together. Or they may encounter each other every Sunday in church. In the case of the bats, they hang about the same cave or hollow. In that way, they can form a "contract" based on helping each other.

The bats are one often cited example of direct reciprocity in nature. Another can be found on coral reefs, where fish of all kinds visit "cleaning stations" where they are scrubbed of parasites by smaller varieties of fish and by shrimps: the former get cleaned of pesky parasites and the latter get a free meal. When a wrasse tends a great grouper, the little cleaner sometimes swims into the gill chambers and mouth, demonstrating remarkable faith that it is not going to be eaten. When the grouper wants to depart, it tells its cleaner that it wants to go by closing its mouth a little and shaking its body. It does this even when it is in danger of being attacked. A safer way to proceed would be to gulp down the cleaner and leave immediately. The first strategy would be a form of cooperation, the second a form of defection.

The nuisance of parasites—ticks—has led to the emergence of another instance of this mechanism at work, in the form of reciprocal grooming, this time among impala, a kind of antelope found in Africa. And when it comes to our closest relatives, textbooks are crammed with examples. One was reported in 1977 by Craig Packer in the Gombe Stream Research Centre, Tanzania, where there has been a long-term study of olive baboons, so named because of their distinctive fur. Packer, now at the University of Minnesota, reported how one male will help another who had previously come to his aid in ganging up on more senior baboons, so that one of them can have sex with the senior's female. Even though the helper will not have sex immediately after forming a coalition, he still cooperates because he expects the favor will be returned.

Sri Lankan macaques *Macaca sinica* will tend the wounds of a fellow male in order to secure the latter macaque's support in future conflicts. Unsurprisingly, juvenile males are especially attentive to the injuries of hefty adults, who can provide more muscle in a future fracas. One study of macaques in Kalimantan Tengah, Indonesia, went so far as to suggest that males were more likely to mate with females that they have previously groomed, the grooming being a kind of payment for sex, a finding given the colorful interpretation that the "oldest profession"—prostitution—seems to date back long before humans.

Male chimpanzees share meat to bind social alliances, and there is some evidence that they increase the degree to which they cooperate in line with how much a partner has been helpful toward them. Reciprocity can be exchanged in all kinds of currencies, such as grooming, support in fights, babysitting, warning, teaching, sex, and of course food. Frans de Waal of Emory University, Atlanta, observed how a top male chimpanzee, Socko, had more chance of obtaining a treat from his fellow chimp May if he had groomed her earlier that day.

There are caveats, however. One is that different scientists use terms such as reciprocity in various ways. Another is that, when it comes to observing behaviors in the wild, it can take many lengthy and detailed studies to understand what is really going on. Tim Clutton-Brock, a professor of ecology and evolutionary biology at Cambridge University, says that it can be hard to sift concrete examples of reciprocity from the illusory ones that can be explained another way.

Let's take Craig Packer's inspirational olive baboon research, for example. Packer had originally thought that the males were trading favors in their pursuit of sex. His original argument went that the allies switch roles, so that each one benefits from the association. But follow-up studies suggested that the cooperating males actually compete with each other when it comes to snatching the prize. The only way they can have an opportunity to mate is to join forces and to cooperate, true enough. But once the existing consort is driven off, then it is every man for himself when it comes to getting the girl. Packer puts it like this: "In this scenario, cooperation is like a lottery, and you can't win if you don't buy a ticket. Because two against one gives very good odds of success, the price of the ticket is very low compared to the value of the prize. Participate in enough lotteries of this sort, and you will always come out ahead—and so will your partners."

RECIPROCITY RULES

> Oliver: *I remember you!*
>
> Grocer: *And I remember you too. Now get out of my store and stay out!*
>
> Oliver: *Oh, don't be like that. Let bygones be bygones. Let's help each other. You have a business, and we have a business. We'll send people to your store, and you send people to our store. What do you say?*
>
> Grocer: *You mind your business and I'll mind my business. Now get out before I throw you out!*
>
> —Laurel and Hardy in *Tit for Tat*

One way to determine which examples of direct reciprocity are real is to think about the qualities that are necessary for this mechanism to work. The evolution of cooperation by direct reciprocity requires that players recognize their present partner and remember the outcome of previous encounters with him or her. They need some memory to remember what another creature has done to them, and a little bit of brainpower to figure out whether to reciprocate. In other words, direct reciprocity requires reasonably advanced cognitive abilities.

I am sure that enough cognitive capacity is available in certain species of birds and among our closer relatives, most certainly the great apes. I am certain there is enough grey matter when it comes to human beings. If Harry does Fred a favor, Fred can remember what Harry looks like. He can also remember his good deed and how Harry has behaved in the past. Fred certainly has sufficient cognitive capacity to figure out from what he can remember if Harry is trustworthy and then tailor his behavior accordingly.

When it comes to the soap opera of everyday life, examples of direct reciprocity are everywhere. The running of a household depends on a ceaseless, mostly unconscious bartering of goods and services. In the kitchen, the one who cooks is often spared the drudgery of the washing up and vice versa. The concord among the members of a student house depends on everyone contributing equitably to cleaning duties, a food kitty, or whatever. If a friend helps us to move house, there is an obligation on us to help to pack his furniture when it is time for him to move, or unpack his crates. Families often harbor expectations that children will reciprocate for the care they receive as babies and as children by looking after their elderly parents.

When we receive invitations to dinner, a night at the theater, and so on, there comes an unwritten obligation to reciprocate in some way, in kind or with a treat in return. If a colleague at work hands you a gift-wrapped present, you make a mental note to reciprocate when her birthday comes around. When someone holds open a door, or gestures toward the mountain of food in a buffet, and says, "After you," many instantly reply, "No, you go first." The same sense of duty to reciprocate helps to make the ritual gift giving at Christmas expensive. And it can be found in bigger tribes and groups of people: businesses may have long-term contractual obligations with each other; governments make treaties with one another; and so on and so forth.

We repay meanness in the same coin. This is best reflected in the phrase "an eye for an eye, a tooth for a tooth," from Exodus 21:24–

27, in which a person who has taken the eye of another in a fight is instructed to give equitable recompense—his own. In the code of Hammurabi, created by an ancient Babylonian king, the principle of reciprocity is expressed in exactly the same way ("If a man put out the eye of another man, his eye shall be put out" and "If a man knocks the teeth out of another man, his own teeth will be knocked out"). One can see the same tit-for-tat logic in the idea of a "just war," where the methods used to prosecute a conflict are proportionate to a given threat.

Unsurprisingly, given its central role in human life, reciprocity has inspired comedy. The vintage duo Laurel and Hardy used acts of slapstick revenge to give their movies a satisfying climax. One of their short films released in 1935 revolves entirely around reciprocal retaliations. Appropriately enough, the film is titled *Tit for Tat*.

So there's plenty of evidence that we live in a reciprocating world. But, of course, it does not always follow that another player in the game of life will reciprocate. Because there is a cost involved in helping another, cooperation always comes with the threat of exploitation. Why should anyone share in hard work or return a favor? Why not cheat? Why not let the other guy toil and sweat, so you can reap the rewards of his hard work and not bother to do a similar favor? In fact, why do we bother with helping others at all?

After all, natural selection puts a premium on passing genes to future generations, and how can it shape a behavior that is "altruistic" in the long term when defection offers such tempting short-term rewards? In modern society, a hefty apparatus of law and order ensures that this temptation to cheat will remain, in general, resistible. But how can direct reciprocity work in the absence of authoritarian institutions? Why, in the case of cleaning stations on the reef, do clients refrain from eating their helpful cleaners after the little fish have discharged their duties?

This issue has been discussed for decades but, from the perspective of my field, was first framed the right way in a paper by Robert Trivers, an American evolutionary biologist. A fascinating character, Trivers, who suffers from bipolar disorder, became steeped in controversy because of his friendship with the leader of the Black Panther Party,

Huey Newton. Today, at Rutgers, the State University of New Jersey, he specializes in the study of symmetry in human beings, "especially Jamaican." Steven Pinker hails Trivers as one of the greats in western intellectual history.

One of the reasons Pinker rates him so highly is a milestone paper that Trivers published in *The Quarterly Review of Biology* in 1971, inspired by a visit to Africa, where he had studied baboons. In "The Evolution of Reciprocal Altruism" Trivers highlighted the conundrum of cheats by borrowing a well-known metaphor from game theory. He showed how the conflict between what is beneficial from an individual's point of view and what is beneficial from the collective's point of view can be encapsulated in the Prisoner's Dilemma. As I explained in the last chapter, it is a powerful mathematical metaphor to sum up how defection can undermine cooperation.

At that time, Trivers did not refer to direct reciprocity but used the term "reciprocal altruism," where altruism is an unselfish concern for the welfare of others. Although altruism is the opposite of the "selfish" behavior that underpins the more traditional view of evolution, it comes loaded with baggage when it comes to underlying motive. Over the course of this book I hope it will become clear that, although it seems paradoxical, "altruistic" behavior can emerge as a direct consequence of the "selfish" motives of a rational player.

Among the mechanisms to escape from the clutches of the Prisoner's Dilemma, the most obvious one, as I have already hinted, is simply to repeat the game. That is why cooperation by direct reciprocity works best within a long-lived community. In many sorts of society, the same two individuals have an opportunity to interact not once but frequently in the village pub, workplace, or indeed the coral reef. A person will think twice about defecting if it makes his co-player decide to defect on the next occasion, and vice versa. The same goes for a fish.

Trivers was the first to establish the importance of the repeated—also known as the iterated—Prisoner's Dilemma for biology, so that in a series of encounters between animals, cooperation is able to emerge. He cited examples such as the cleaner fish and the warning cries of birds. What is remarkable is that Trivers went further than this. He discussed how "each individual human is seen as possessing altruistic and cheating tendencies," from sympathy and trust to dishonesty and hypocrisy.

Trivers went on to suggest that a large proportion of human emotion and experience—such as gratitude, sympathy, guilt, trust, friendship, and moral outrage—grew out of the same sort of simple reciprocal tit-for-tat logic that governed the daily interactions between big fish and the smaller marine life that scrubbed their gills. These efforts built on earlier attempts to explain how reciprocity drives social behavior. In the *Nicomachean Ethics,* Aristotle discusses how the best form of friendship involves a relationship between equals—one in which a genuinely reciprocal relationship is possible. In Plato's *Crito,* Socrates considers whether citizens might have a duty of gratitude to obey the laws of the state, in much the way they have duties of gratitude to their parents for their existence, sustenance, and education. Overall, one fact shines through: reciprocity rules.

THE ITERATED DILEMMA

Since the Prisoner's Dilemma was first formulated in 1950, it has been expressed in many shapes, forms, and guises. The game had been played in a repeated form before, but Trivers made a new advance when he introduced the repeated game to an analysis of animal behavior. This iterative Prisoner's Dilemma is possible in a colony of vampire bats and at the cleaning stations used by fish on a reef, which were the subject of Trivers's paper.

However, the implications of what happens when the Prisoner's Dilemma is played over and over again were first described before Trivers's analysis in 1965 by a smart double act: Albert Chammah, who had emigrated from Syria to the United States to study industrial engineering, and Anatol Rapoport, a remarkable Russian-born mathematician-psychologist who used game theory to explore the limits of purely rational thinking and would come to dedicate himself to the cause of global peace. In their book, *Prisoner's Dilemma,* they gave an account of the many experiments in which the game had been played.

Around the time that Trivers made his contribution, another key insight into the game had come from the Israeli mathematician Robert J. Aumann, who had advised on cold war arms control negotiations in the 1960s and would go on to share the Nobel Prize in Economics in 2005. Aumann had analyzed the outcome of repeated encounters and demonstrated the prerequisites for cooperation in various situations—for instance, where there are many participants, when there is infrequent interaction, and when participants' actions lack transparency.

In the single shot game, the one that I analyzed earlier in the discussion of the payoff matrix of the Prisoner's Dilemma, it was logical to defect. But Aumann showed that peaceful cooperation can emerge in a repeated game, even when the players have strong short-term conflicting interests. One player will collaborate with another because he knows that if he is cheated today, he can go on to punish the cheat tomorrow. It seemed that prospect of vengeful retaliation paves the way for amicable cooperation. By this view, cooperation can emerge out of nothing more than the rational calculation of self-interest. Aumann named this insight the "folk theorem"—one that had circulated by word of mouth and, like so many folk songs, has no original author and has been embellished by many people. In 1959, he generalized it to games between many players, some of whom might gang up on the rest.

This theorem, though powerful, does not tell you how to play the game when it is repeated. The folk theorem says there is a strategy that can induce a rational opponent to cooperate, but it does not say what is a good strategy and what is a bad one. So, for example, it could show that cooperation is a good response to the Grim strategy. That strategy says that I will cooperate as long as you cooperate, but if you defect once then I will permanently switch to defection. In reality, such strategies are far from being the best way to stimulate cooperation in long-drawn-out games.

To find out how to play the game, thinkers in the field had to wait for a novel kind of tournament, one that would shed light on all the nuances of the repeated Prisoner's Dilemma. This was developed by Robert Axelrod, a political scientist at the University of Michigan, who turned the results into a remarkable book, *The Evolution of Cooperation*, which opens with the arresting line "Under what conditions will cooperation emerge in a world of egoists without central authority?" In his direct prose, Axelrod clearly described how he had devised a brilliant new way to tease out the intricacies of the Dilemma.

He organized an unusual experiment, a virtual tournament in a computer. The "contestants" were programs submitted by scientists so they could be pitted against each other in repeated round-robin Prisoner's Dilemma tournaments. This was the late 1970s and at that time the idea was breathtakingly novel. To put his tournaments in context—commercial, coin-operated video games had only appeared that same decade. But Axelrod's idea was no arcade gimmick. Unlike humans, who get bored, computers can tirelessly play these strategies against each other and unbendingly stick to the rules.

Researchers around the world mailed Axelrod fourteen different programs. He added one of his own—one that randomly cooperates and defects—and pitched all of them against each other in a round-robin tournament. Success was easy to measure. The winner would be the strategy that received the highest number of points after having played all other strategies in the computer over two hundred moves. During the entire tournament, Axelrod explored 120,000 moves and 240,000 choices.

Because the computers allowed for limitless complexity of the programs entered into the tournament, one might expect that the biggest—and thus "smartest"—program would win. But size is not everything. In fact, the simplest contestant won hands down, much to the surprise of the theorists. The champion turned out to be a measly little four line computer program devised by none other than Anatol Rapoport.

Called Tit for Tat, the strategy starts with a cooperative move and then always repeats the co-player's previous move. A player always starts by keeping faith with his partner but from then on mimics the last move of his opponent, betraying only when his partner betrays him. This is more forgiving than Grim, where a single defection triggers an eternity of defection.

Standing back from the Prisoner's Dilemma, it is easy to see the advantage of adopting a simple strategy. If you are too clever, your opponent may find it hard to read your intentions. If you appear too unresponsive or obscure or enigmatic, your adversary has no incentive to cooperate with you. Equally, if a program (or a person for that matter) acts clearly and sends out a signal that it cannot be pushed around, it does make sense to cooperate.

What was also striking was that this discovery was old hat. The contestants in the computer Prisoner's Dilemma tournament already knew about this powerful strategy. Work published at the start of that decade had shown that Tit for Tat does well. Indeed, the strategy carries echoes of the one that the nuclear powers had adopted during the cold war, each promising not to use its stockpiles of A- and H-bombs so long as the other side also refrained. Many of the contestants tried to improve on this basic recipe. "The striking fact is that none of the more complex programs submitted was able to perform as well as the original, simple Tit for Tat," observed Axelrod.

When he looked in detail at the high-ranking and low-ranking strategies to tease out the secret of success, Axelrod found one property in particular appeared to be important. "This is the property of being nice, which is to say never being the first to defect." This strategy is interesting because it does not bear a grudge beyond the immediate retaliation, thereby perpetually furnishing the opportunity of establishing "trust" between opponents: if the opponent is conciliatory, both reap the rewards of cooperation.

Axelrod went on to organize a second tournament, this time attracting sixty-three entries from six countries, ranging from a ten-year-old computer obsessive to a gaggle of professors of various persuasions. One entry arrived from the British biologist John Maynard Smith, whom we will learn much more about later. Maynard Smith submitted Tit for Two Tats, a strategy that cooperates unless the opponent has defected twice in a row. Maynard Smith, a revered figure in his field, limped in at twenty-fourth place.

Rapoport, however, followed the maxim of British soccer leagues: "Never change a winning team." Once more, he fielded the Tit-for-Tat strategy, and once again it won: it really did pay to follow this simple strategy. This was the very tournament that had inspired Karl Sigmund to focus on the Dilemma and that, in turn, inspired me when he gave me that sermon on the mountain. Robert Axelrod's book *The Evolution of Cooperation* is now regarded as a classic in the field, and deservedly so.

But did Axelrod's computer tournament have anything to say about the real world? Yes. A real-life example of such a contest was reported in 1987, when Manfred Milinski, now the director of the Max Planck Institute for Evolutionary Biology in Ploen, Germany, studied the behavior of stickleback fish. When a big predator such as a pike looms, one or more of a school of sticklebacks will approach to see how dangerous it is. This "predator inspection" is risky for these scouts, but the information can benefit them as well as the rest of the school—if the interloper is not a predator or if it has just fed and is not hungry, the smaller fish don't need to move away. Assessing whether it is necessary to flee seems foolish but is important because in their natural habitat there are many pike and other fish swimming about, so moving away is not always a good strategy: one can jump out of the way of one snapping predator into the jaws of another.

Milinski found that stickleback fish rely on the Tit-for-Tat strategy during this risky maneuver. If a pike shows up in the neighborhood, two sticklebacks often swim together in short spurts toward the open mouth of the predator to size him up. Each spurt can be thought of as a single round of the Dilemma. Cooperating in this game of chicken is best for both fish, since it cuts the risk of being eaten. This is due to the "predator confusion" effect: pike can waste valuable time when they have to decide which of two or more prey to strike first, a real-life version of the paradox of Buridan's ass, the hypothetical situation in which a donkey cannot choose between two stacks of hay and so dies of hunger. Yet each little fish has an understandable incentive to hang back a little and let the other stickleback soak up more of the risk.

To investigate what was going through their little fishy heads, Milinski made ingenious use of a mirror. When held in the right place, it could create the illusion that a single stickleback was accompanied by another scout. By tilting the looking glass, Milinski could make it seem to a stickleback scout that his mirror-image "companion" was either cooperating by swimming alongside or falling behind and defecting, like the officer leading the charge who slowly slips behind his troops and out of harm's way. The lead scout would often react to the apparent defection of its mirror fish by slowing down or turning tail, without completing its scouting mission. If the mirror image kept pace with the scout, the latter usually approached the predator more closely than it would if swimming alone.

NOISE

So far, so satisfyingly straightforward. But there is a problem with Tit for Tat, one that is not immediately obvious when using computer programs that interact flawlessly. Humans and other animals make mistakes. Sometimes their wires get crossed. Sometimes the players become distracted. They suffer mood swings. Or they simply have a bad day. Nobody's perfect, after all. One type of mistake is due to a "trembling hand": I would like to cooperate but I slip up and fail to do so. Another is caused by a "fuzzy mind": I am convinced that this person was mean to me and defected in the last round, when in fact he did not. Perhaps I was confusing him with someone else. Trembling hands and fuzzy minds lead to what I call "noisy" interactions.

The significant role of noise for the evolution of cooperation was first pointed out in a paper in the journal *Nature* by Robert May of Oxford University, a brilliant former physicist who would come to exert a profound influence on theoretical biology. Bob (being Australian, he prefers "Bob") is best known for the great strides he made in putting ecology on a mathematical basis. In his short essay he argued that evolutionary biologists should study the influence of mistakes on the repeated Prisoner's Dilemma. He realized that the conclusions from a game that is perfectly played, as was the case in Axelrod's tournaments, are not necessarily robust or realistic.

This is an important point. Even infrequent mistakes can have devastating consequences. When pitched against another player adopting the same approach, the Tit-for-Tat strategy can trigger endless cycles of retaliation. Since all it knows how to do is strike back at defectors, one scrambled signal or slipup can send Tit for Tat spiraling ever downward into vendettas that overshadow those seen in *Romeo and Juliet,* between the Hatfields and McCoys, or anything witnessed in Corsica, for that matter. The obvious way to end this bloody spiral of retaliation is to let bygones be bygones: for example, only to demand revenge now and again, or to decide it by the throw of a die. Inspired by this important insight, I would extend Axelrod's pioneering work and incorporate the effects of noise to make it more true to life.

TAKE ADVANTAGE OF MISTAKES

As I studied for my doctorate with Karl, we devised a way to take confusion, slips, and mistakes into account. In the jargon, instead of the conventional deterministic strategies we used probabilistic strategies, where the outcome of the game becomes more fuzzy and random. We decided to explore the evolution of cooperation when there is noise by holding a probabilistic tournament in a computer, building on Axelrod's pioneering work. The idea was to use a spectrum of strategies, generated at random by mutation and evaluated by natural selection.

All of our strategies were influenced by chance. They would cooperate with a certain probability after the opponent had cooperated and they would also cooperate with a certain probability after the opponent had defected. Think of it this way: we are able to put varying shades of "forgiveness" in the set of strategies that we explore. Some forgive one out of two times. Others one out of five defections, and so on. And some strategies, of course, are unbending. These Old Testament style strategies almost never forgive. As was the case with the Grim strategy, they refuse ever to cooperate again after an opponent has defected only once.

To study the evolution of cooperation, we seasoned the mix with the process of natural selection so that winning strategies multiplied while less successful rivals fell by the wayside and perished. The strategies that got the most points would be rewarded with offspring: more versions of themselves, all of which would take part in the next round. Equally, those that did badly were killed off. For extra realism, we arranged it so that reproduction was not perfect. Sometimes mutation could seed new strategies.

Now Karl and I could sit back and watch the strategies slug it out in our creation over thousands and thousands of generations. Our fervent hope was that one strategy would emerge victorious. Even though no evolutionary trajectory ever quite repeated itself, there were overall patterns and consistency in what we observed. The tournament always began with a state of "primordial chaos." By this I mean that there were just random strategies. Out of this mess, one, Always Defect, would inevitably take an early lead: as is so often seen in many Hollywood movies, the baddies get off to a flying start.

For one hundred generations or so, the Always Defect strategy dominated our tournament. The plot of life seemed to have a depressing preface in which nature appeared cold-eyed and uncooperative. But there was one glimmer of hope. In the face of this unrelenting enemy, a beleaguered minority of Tit for Tat players clung on at the edge of extinction. Like any Hollywood hero, their time in the sun would eventually come: when the exploiters were left with no one left to exploit, and all the suckers had been wiped out, the game would suddenly reverse direction. Karl and I took great pleasure in watching the Always Defectors weaken and then die out, clearing a way for the triumphant rise of cooperation.

When thrown into a holdout of die-hard defectors, a solitary Tit for Tat will do less well than defecting rotters, because it has to learn the hard way, always losing the first round, before switching into retaliatory mode. But when playing other Tit for Tat–ers, it will do significantly better than Always Defect and other inveterate hard-liners. In a mixture of players who adopt Always Defect and Tit for Tat, even if the latter only makes up a small percentage of the population, the "nice" policy will start multiplying and quickly take over the game. Often the defectors do so poorly that they eventually die out, leaving behind a cooperative population consisting entirely of Tit for Tat.

But Karl and I were in for a surprise. In our computer tournaments,

Tit for Tat–ers did not ultimately inherit the Earth. They eventually lost out to their nicer cousins, who exploited Tit for Tat's fatal flaw of not being forgiving enough to stomach the occasional mishap. After a few generations, evolution will settle on yet another strategy, which we called Generous Tit for Tat, where natural selection has tuned the optimum level of forgiveness: always meet cooperation with cooperation, and when facing defection, cooperate for one in every three encounters (the precise details actually depend on the value of the payoffs being used). So as not to let your opponent know exactly when you were going to be nice, which would be a mistake (John Maynard Smith's Tit for Two Tats strategy could be easily exploited by alternating cooperation and defection), the recipe for forgiveness was probabilistic, so that the prospect of letting bygones be bygones after a bad move was a matter of chance, not a certainty. Generous Tit for Tat works in this way: never forget a good turn, but occasionally forgive a bad one.

Generous Tit for Tat can easily wipe out Tit for Tat and defend itself against being exploited by defectors. The Generous strategy dominates for a very long time. But, due to the randomness in our tournaments, it does not rule forever. We observed how slowly, almost imperceptibly, a population of Generous Tit for Tat mutates and drifts toward more and more lenient strategies. Ultimately, the population becomes uniformly nice: all cooperate. The reason is that when everybody tries to be nice, forgiveness pays handsomely. There is always an incentive to forgive quicker and quicker because the highest rewards come from having many productive (that is, cooperative) interactions. Now, at last, defectors have a chance to rise up again, with the help of the right mutation. A population of nice players who always cooperate is dry tinder for an invasion by any lingering or newly emerged defector. In this way, the cycle starts anew.

These probabilistic games are always different in detail. But there was a pattern overall. Karl and I would always see the same strategies wax and others wane. Overall, the cycles play out in a predictable way, sweeping from all defectors to Tit for Tat, to Generous Tit for Tat, then all cooperators. Finally, with a great crash, the makeup of the community lurches back to being dominated by dastardly defectors all over again.

The good news is that a reasonably nice strategy dominates the tournament. When you average out the strategies over the entire duration of a game, the most common is Generous Tit for Tat. The bad news is that, in the real world, these cycles could sweep out over years, decades, or even centuries. Plenty of anecdotal evidence suggests that these cycles turn in human history too. Kingdoms come and go. Empires spread, decline, and crumble into a dark age. Companies rise up to dominate a market and then fragment and splinter away again in the face of thrusting, innovative competitors.

Just as these tournaments never see one strategy emerge with total victory, so it seems that a mix of cooperators (law-abiding citizens) and defectors (criminals) will always persist in human societies. The same goes for beliefs. One faith rises and another declines, the very scenario that prompted Augustine to write *The City of God* (*De civitate Dei*) after Rome was sacked by the Visigoths in 410. Augustine wanted to counter claims that Rome had been weakened by adopting Christianity, but as our computer tournaments made clear, great empires are destined to decline and fall: it was more a case of *delapsus resurgam*—when I fall I shall rise—and vice versa.

As the latest recession has vividly underlined, and as has been noted over the past few decades, there are economic cycles too. Regulations are introduced, then people figure out clever ways to evade them over the years. Periods of hard work and grinding toil are followed by those of leniency, when people slacken, take time off, and exploit the system. In our computer simulations, had we stumbled upon a mathematical explanation for the fundamental cycles of life that endlessly whirl around phases of cooperation and defection?

GOODBYE VIENNA

After four papers and a little more than one year of collaboration, Karl told me that I had done enough research to complete my thesis on the evolution of cooperation. I immediately got on with typing up my work. A few days later, I handed him my thesis. He held it up and closely examined the slender document from the side, shook his head, and declared that it was too thin: "A PhD thesis has to be thicker." The next day I gave him the same thesis, only this time the font was bigger and the line spacing doubled. Karl was not fooled. But he was a pragmatist. He looked at it once again and said: "That's all right."

Karl suggested that I apply for a position with the leading figure in our field, Bob May at the University of Oxford. At that time, Bob was famous for the way in which he had injected the rigor of math into biology to reveal the underlying order in the living world. He had studied whether stability is the cause of the diversity of ecosystems, or whether it is the other way round (it turns out that populating an ecosystem with a diverse range of living things does not automatically mean stability). He charted the relationships between insects and their parasites. Using mathematical models, Bob had revealed how connections between species could lead to fluctuations in the number of individuals. In this way Bob had introduced chaos into biology—revealing how *apparently* random and complex behavior is ordered by simple underlying rules (I am writing this at home while sitting at the very same desk at which Bob made this discovery—a gift from him to help furnish my first house).

Karl did not rate my chances of moving to Oxford very highly, so I had also applied to Berkeley and Göttingen. My future life, career, and everything now depended on insubstantial aerograms. As these air mail letters winged their way around the world my predicament was both romantic and sad. I was about to marry Ursula and our time in Vienna was drawing to a close. The melancholy of leaving home was tempered by the excitement of a new adventure. Neither of us knew where on the planet we would end up.

Initially, Karl's judgment seemed spot on. Bob rejected me, saying he did not have a group. Nor did he work much with postdoctoral students. I wrote to him again, pointing out that I might bring along my own funding, an Erwin Schrödinger Fellowship. By this time, Karl was lobbying Bob too. Eventually, to my delight, he agreed. At last the next step in my career was clear—up to a point. I had absolutely no idea what to expect at Oxford.

Ursula and I got married in Vienna the month before our move. We said our goodbyes after the service and went home to our respective parents until the time came to catch a train for what would turn out to be a nine-year honeymoon, starting in 1989. The day of departure saw us laden with seven suitcases and two bikes. It was cold and windy. A battleship grey sky threatened a torrential downpour. Our families saw us off that night from the Westbahnhof in Vienna. A friend stiffly stood before me and formally shook my hand. "Don't embarrass us," he joked. As the train pulled out into the darkness, my new wife cried.

The next day, once the cross-Channel ferry had set us down, I caught my first glimpse of Britain. It was not William Blake's green and pleasant land. The soil was cracked and dry. The grass and foliage were brown and the country was in the grip of a drought. Reservoirs were drained and there were hosepipe bans and fines for anyone found washing a car. In Plymouth, flower beds were being showered with treated sewage effluent. In one British zoo, dirty water from the penguin pool was being sprinkled on parched putting greens. As our train waited, a fire was put out on the tracks ahead.

My expectations took a sharp departure from reality once again, when I eventually walked into my new place of work, the Zoology Department at the University of Oxford, an unlovely concrete pile on South Parks Road. There were posters showing birds and other animals. But there was not an equation or diagram in sight. Was I in the right place? I was. And I would discover that I was lucky to be there. There was little in the way of formalities. Unlike the hierarchical Austrian academic system, which discourages lowly students from bothering busy Herr Professors, I found myself having an informal chat over a cup of coffee or afternoon tea with many influential figures, from the great Bill Hamilton, who did pioneering work on cooperation, to Sir Richard Southwood, Richard Dawkins, Paul Harvey, and John (later, Lord) Krebs. This was a wonderful, heady intellectual atmosphere. I fell in love with the place.

Bob May would sometimes play soccer with everybody—all the students and professors were as obsessed by games as I was. This was a worry, given that he was so intensely competitive. In the British tradition, winning was beside the point and taking soccer too seriously was frowned upon. But not when it came to this wiry, quick Australian. Fortunately for the rest of us, he was somewhat ineffective. Appropriately enough, the goddess of randomness did smile on him from time to time, however. During one early encounter, when the score was seven all and I was the goalkeeper, Bob kicked the ball past me in the very last minute of the game. Jubilant, Bob screamed: "Martin, this was excellent for your career!"

He and I are so different, the odd couple. He is a compact, frizzy haired wisecracker who has little sympathy for religion. I tower over him, a balding Catholic with a Schwarzenegger-like English accent that is a gift when it comes to recording the message on telephone answering machines ("I'm away at the moment but I'll be back!"). Bob is endowed with a heady blend of traits: a passion for precision, an equal love of profanity, and a hilarious disdain for his peers ("A biologist is someone who wanted to be a scientist but was not good enough to be a physicist"). We were united by our love of games, from the mathematical to the physical, and we both wanted to win. He was bemused when I told him that, remarkably enough, my German lacks an everyday word for "competitive."

Our rapport had an energizing effect on my work. For my first project, I followed up an idea that first came to me at a high-powered gathering organized by the German Nobel laureate Manfred Eigen in Klosters, Switzerland. During a talk there by Bill Haseltine on the human immunodeficiency virus, HIV, I realized that the body of an AIDS victim must harbor a swarm of closely related replicating viruses. This reminded me of my work with Peter Schuster in mathematical biology. One day, I thought to myself, I would like to develop a mathematical model of virus infections. But my respect at that time for the difficulty of solving problems was almost paralyzing.

I was fortunate that Bob had already studied the virus with a colleague, Roy Anderson. Together, they had charted how the virus spreads between people. But I wanted to take this approach in a new direction. I wanted to model what happens *inside* a person who is unfortunate enough to have been infected with the virus. That would require explaining how the virus spreads between cells in the face of attacks from the body's immune system. To find out how HIV fares in the human body, I would have to use a brand of mathematics similar to that used in my tournaments with Karl.

I discovered that I could explain the puzzlingly long delay between HIV infection and AIDS and why this period can vary so greatly between patients—it could show up in less than two years in one person and yet lurk for more than a decade in another. What was remarkable was that I could draw my conclusions from existing data without the need for new experiments on animals or trials on patients. All I needed was a ready supply of computer number crunching power to explore the way that the virus breeds and mutates inside the body.

Bob was so thrilled by this result that he insisted that I show my findings to Roy Anderson, who was by then working at Imperial College London. He too was amazed. After I published the first results in the journal *AIDS* in 1990, an extended version of my theory and clinical data came out in the prestigious journal *Science* the year after. I worked, too, on hepatitis B virus with Barry Blumberg, master of Balliol College, who won the Nobel Prize for discovering the virus and making a vaccine. This kind of research helped to establish the field that is now known as virus dynamics, where mathematical models chart out the progress of virus infections within infected hosts.

SOARING EAGLES, DIVING STRATEGIES

Karl and I had so many games left to play, with so many variants and so many potential outcomes. In 1992, our work on Generous Tit for Tat was published in the British journal *Nature,* which shares with the American journal *Science* the distinction of being the journal that scientists want to appear in most of all. Karl and I had plenty of new ideas when it came to extending our work. My second summer at Oxford, I once again returned to Austria to resume our explorations of the Prisoner's Dilemma.

Previously, Karl and I had calculated the strategies that emerge when the decision of a player only depended on the opponent's last move.

But, of course, this only gives a partial picture of what can happen. We now wanted to look at strategies that also take into account the player's own moves. Let me give you an example to show exactly what I mean by this. Put yourself in the position of a contestant in one of our tournaments. You might be less annoyed with a fellow player who had defected if you had defected too. Equally, you might be more angry with him if you had cooperated.

To find out whether this influenced the winning strategies, I found myself with my new portable computer and Karl in a room of Rosenburg Castle, a fabulous medieval heap in lower Austria, complete with an arcaded yard that was once used for jousting. I was working in this fairy-tale setting because I had to be with Karl. And Karl was there because he had to be with his wife, who was staying in Rosenburg to carry out research on the historical building.

Although I did not know what would happen in the new computer experiments, I had a pretty good idea. Generous Tit for Tat would win the games again. Simple. As Karl and I went through the motions to show that this was indeed the case, there was only one distraction. The castle has a fine collection of birds of prey that, at preordained times, performed in the big courtyard. Handlers clad in Renaissance garb lured the raptors to make spectacular dives, when they skimmed the heads of spectators. They soared and swooped between various points on the façade as Karl and I looked on.

We ran our simulations again and again, taking a break every so often to watch the highlight of these displays, a thousand foot dive by a golden eagle. By then these magnificent birds were a welcome distraction because there was a problem. My favorite, the Generous Tit for Tat strategy, was being beaten again and again in the jousting tournaments on my laptop. This was frustrating, since I had confidently expected this strategy to rule the roost. I found myself wishing that there were more birds to take my mind off my work. There had to be a bug in my program. I checked. And I checked again. I couldn't find it. I pride myself on being able to do this and had a fail-safe rule: "the bug is always where you are not looking." Finally the simple truth dawned on me. There wasn't a glitch this time.

The losing streak of Generous Tit for Tat was telling me something important but at that particular moment I wasn't listening. I hunted for a way to make the problem go away. But I could not save Tit for Tat. After a few days, having reluctantly decided that the result might be real, I took a closer look and found that a new strategy consistently won. It consisted of the following instructions and they seemed bizarre at first glance:

If we have both cooperated in the last round, then I will cooperate once again.

If we have both defected, then I will cooperate (with a certain probability).

If you have cooperated and I have defected, then I will defect again.

If you have defected and I have cooperated, then I will defect.

Overall, this means whenever we have done the same, then I will cooperate; whenever we have done something different, then I will defect. In other words, this winning strategy does the following: If I am doing well, I will repeat my move. If I am doing badly, I will change what I am doing. I now became intrigued. My mood lifted.

Back in Oxford I told the distinguished biologist John Krebs about the winning strategy when I bumped into him in the corridor of the Zoology Department. He recognized it instantly. "This sounds like Win Stay, Lose Shift, a strategy which is often considered by animal behavioralists." The strategy was much loved by pigeons, rats, mice, and monkeys. It was used to train horses too. And it had been studied for a century. John was amazed at how the strategy had evolved by itself in a simple and idealized computer simulation of cooperation. So was I.

Now I had to figure out why Win Stay, Lose Shift was better than either Tit for Tat or Generous Tit for Tat. The answer was revealed by studying the details of the cycles of cooperation and defection that turned in my laptop. In the earlier work, one can mark the end of one cycle and the start of the next by the emergence of a population of unconditional cooperators. With random mutations thrown in the mix, a defector always emerges to take over that docile population, marking the start of a new cycle. I discovered that the secret of Win Stay, Lose Shift lay at this stage, when cooperation is at a peak and nice strategies abound. It turns out that unconditional cooperators can undermine the strategies of Tit for Tat and Generous Tit for Tat. But they can't beat Win Stay, Lose Shift.

In a game with some realistic randomness, Win Stay, Lose Shift discovers that mindless (or unconditional) cooperators can be exploited. The reason is easy to understand: any little mistake will reveal that such a cooperator will still carry on being nice in the face of nastiness. And, as the name suggests, Win Stay, Lose Shift carries on exploiting its fellow players, when it is not punished with retaliation. Or, as Karl and I put it, this strategy cannot be subverted by softies. This characteristic turns out to be an important ingredient of its success.

The deeper lesson here is that a strategy that does not appear to make sense when played in a straightforward deterministic way can triumph when the game of life is spiced up with a little realistic randomness. When we surveyed the existing literature, it turned out that others had studied the very same strategy, under various guises. The great Rapoport had dismissed it, calling it "Simpleton" because it seemed so stupid—in encountering a defector, it will alternate between cooperation and defection. He reasoned that only a dumb strategy would cooperate with a defector every other time.

But the strategy is, in fact, no simpleton. Our work made it clear that randomness was the key to its success. When confronted with defectors, it would cooperate unpredictably, with a given probability, protecting it from being exploited by opportunists. The same strategy was called "Pavlov" by David and Vivian Kraines of Duke University and Meredith College, North Carolina, who had noted that it could be effective. Moreover, two distinguished American economists, Eric Maskin and Drew Fudenberg, had also shown that such a strategy can achieve a certain level of evolutionary stability for about half of all Prisoner's Dilemmas. But they had all looked at a deterministic (nonrandom) version of Win Stay, Lose Shift, when it was the probabilistic version that was the winner in our Rosenburg tournaments.

In the great game of evolution, Karl and I found that Win Stay, Lose

Shift is the clear winner. It is not the first cooperative strategy to invade defective societies but it can get a foothold once some level of cooperation has been established. Nor does it stay forever. Like Generous Tit for Tat, Win Stay, Lose Shift can also become undermined and, eventually, replaced. There are and always will be more cycles.

Many people still think that the repeated Prisoner's Dilemma is a story of Tit for Tat, but, by all measures of success, Win Stay, Lose Shift is the better strategy. Win Stay, Lose Shift is even simpler than Generous Tit for Tat: it sticks with its current choice whenever it is doing well and switches otherwise. It does not have to interpret and remember the opponent's move. All it has to do is monitor its own payoff and make sure that it stays ahead in the game. Thus one would expect that, by requiring fewer cognitive skills, it will be more ubiquitous. And, indeed, Win Stay, Lose Shift was a better fit for Milinski's stickleback data than Tit for Tat had been.

In the context of the Prisoner's Dilemma, think of it like this. If you have defected and the other player has cooperated, then your payoff is high. You are very happy, and so you repeat your move, therefore defecting again in the next round. However, if you have cooperated and the other player has defected, then you have been exploited. You become morose and, as a result, you switch to another move. You have cooperated in the past, but now you are going to defect. Our earlier experiments had shown that Tit for Tat is the catalyst for the evolution of cooperation. Now we could see that Win Stay, Lose Shift is the ultimate destination.

Does that mean we had solved the Dilemma? Far from it. Karl and I realized in 1994 that there is yet another facet to this most subtle of simple games. The entire research literature was based on an apparently innocent and straightforward assumption: when two players decide to cooperate or to defect, they do so simultaneously. What I mean by this is that the conventional formulation of the Prisoner's Dilemma is a bit like that childhood game, Rock Scissors Paper. Both players make their choice at precisely the same time.

Karl and I thought that this restriction was artificial. We could think of examples, such as the vampire bats that donate excess blood to hungry fellow bats and creatures that groom each other and so on, where cooperation does not happen simultaneously and partners have to take turns. So we decided to play a variant of the Prisoner's Dilemma, called the Alternating Prisoner's Dilemma, to see if it this change had any effect.

When we played the alternating game we were reassured to find as before that there was a tendency to evolve toward cooperation. We also observed the same cycles that saw the rise, and the fall, of cooperative and defective societies as we had seen in the simultaneous game. Once again, cooperation can thrive. But there was an important twist. We were surprised to find that the Win Stay, Lose Shift principle that had trumped all comers in the simultaneous games (eventually) no longer emerged as victor. Instead, it was Generous Tit for Tat that reigned supreme.

Drew Fudenberg, now a colleague at Harvard, pointed out to me years later that one can think of the alternating and the simultaneous games as two different limiting examples of situations found in the real world. In the alternating game it is your turn and then mine. I get all the relevant information about your move before I need to decide what to do, and vice versa. In the simultaneous game, however, neither of us gets any information about what the other will do in the present round. In everyday life, the reality most likely lies somewhere in between. We might always get some information about what the other person is up to (whether he is delivering his part of the deal or not) but that information may not be complete or reliable.

Manfred Milinski has studied how people use these strategies. In experiments with first-year biology students in Bern, Switzerland, cooperation dominated in both the simultaneous and the alternating Prisoner's Dilemma and he observed how players tended to stick to one strategy, whichever timing of the game they played, with 30 percent adopting a Generous Tit for Tat–like strategy, and 70 percent the Win Stay, Lose Shift. As our simulations had suggested, the latter were more successful in the simultaneous game while Generous Tit for Tat–like players achieved higher payoffs in the alternating game. Both strategies appear to play a role in the ecology of human cooperation.

DILEMMA PAST, DILEMMA FUTURE

Even today, the repeated Prisoner's Dilemma maintains a tight grip on the curious scientist. We have seen how one mechanism to solve the Dilemma and nurture cooperation is direct reciprocity, where there are repeated encounters between two players, whether people, institutions, companies, or countries. At first winning seemed easy with the Tit for Tat strategy, one that at most ends up sharing the wins equally among players. But by adding some randomness, to depict the effect of mistakes, we found that Tit for Tat is too harsh and unforgiving. The strategy triggers bloody vendettas.

We need a sprinkling of forgiveness to get along, and we found it in the strategies of Win Stay, Lose Shift and Generous Tit for Tat. The latter strategy always reminds me of a piece of advice that Bob May once gave me: "You never lose for being too generous." I was impressed by that sentiment because he has thought more deeply about winning and losing than anyone else I know, yet being number one means everything to him. As his wife once kidded, "When he comes home to play with the dog, he plays to win."

Let's compare the successful strategies of Tit for Tat and Win Stay, Lose Shift. Both cooperate after mutual cooperation in the last round. Thus neither is the first to defect, at least intentionally. Only a mistake, a misunderstanding, or simply having a bad day can cause the first defection. If this occurs and the other person defects and I end up being exploited, then both strategies tell me to defect in the next move. If, on the other hand, I defect and the other person cooperates then I switch to cooperation according to Tit for Tat but continue to defect according to Win Stay, Lose Shift.

One can explain the Tit for Tat reasoning as follows: now I feel regret and I want to make up for the defection last round. But the Win Stay, Lose Shift reasoning seems—regrettably—more "human": if we get away with exploiting someone in this round then we continue to do it in future rounds. There's another basic difference between these strategies. If both players defect, then Tit for Tat will also defect and will not attempt to reestablish a good relationship. Win Stay, Lose Shift will cooperate, on the other hand, and try to restore better terms.

Both options make sense, but again it seems that the Win Stay, Lose Shift feels more realistic if we are in a relationship where there is hope of reestablishing cooperation. Overall, Win Stay, Lose Shift can cope better with mistakes because it actively seeks good outcomes, trying to restore cooperation after mutual defection, though it will try to exploit unconditional cooperators. In contrast, Tit for Tat does not exploit unconditional cooperators and does not attempt to resume cooperation after mutual defection.

If one stands back and looks at the way that research on the Prisoner's Dilemma has developed over the years, one key development is the rise in the influence of probabilistic strategies, where the players are likely to play a certain way, and at a certain time, but it is by no means certain they will react in precisely the same way in each and every circumstance. To that we can add another element of realism in acknowledging that real life lies somewhere between playing a simultaneous and an alternating game as a result of the degree to which they alternate and the degree to which they know what the other has done.

These more realistic games also generate cycles, where strategies vary from Always Defect to Tit for Tat to Generous Tit for Tat then to indiscriminate cooperation and then, inevitably, return to square one again as defection takes over. Even though Win Stay, Lose Shift can lengthen the period of cooperation in a cycle, we find that it breaks down eventually to allow a resurgence of defectors.

The cycles that we observe in our tournaments are quite different from the findings of traditional game theory, where there is always an emphasis on stable equilibrium. Without having to go into all the details, this point can be appreciated simply by looking at the language that is used in classical evolutionary and economic game theory. References abound to evolutionarily *stable* strategy and Nash *equilibrium,* for example.

We have moved from old "evolutionary statics" and are now starting to understand the flux and change of "evolutionary dynamics." This classical notion of life evolving to a stable and unchanging state has now been overturned by a much more fluid picture. No strategy is really stable and thus successful for eternity. There is constant turnover. Fortune does not smile forever on one person. A heaven of cooperation will always be followed by a defective hell. Cooperation's success depends on how long it can persist and how often it reemerges to bloom once again. What a fascinating and turbulent insight into the evolution of cooperation and life.

Yet there's still so much more left to find out. We have only explored a small subset of this extraordinary game. There are many more variants out there, huge hinterlands of games that stretch out to a receding horizon. Despite the thousands of papers written on the repeated Prisoner's Dilemma, the mathematical possibilities in this model of direct reciprocity are open-ended, like chess, and not closed, like the strategies for playing tic-tac-toe. Our analysis of how to solve the Dilemma will never be completed. This Dilemma has no end.

CHAPTER 2

Indirect Reciprocity— Power of Reputation

The moment there is suspicion about a person's motives, everything he does becomes tainted.

—Mohandas Gandhi

"Give and it shall be given unto you." This oft-quoted line from Luke's Gospel story of Jesus' birth, ministry, and resurrection seems to be just another example of direct reciprocity, which I described in the last chapter. But take a moment to think about this phrase and you will see that there is a crucial difference: it is not entirely clear who is doing the giving in response to your act of generosity. Perhaps it is a family member, friend, or workmate. But it could be a stranger too, or indeed several strangers.

Many people might interpret the quote as meaning that, if you are generous, a reward is promised in a subsequent world, a paradise or heaven. But my favorite interpretation is that the reward comes to you in the here and now. Kindness will elicit kindness. In this way, circles of humanity, tolerance, and understanding can loop through and around our society. Either way, it is a powerful form of cooperation, and its implications are huge, shaping how we behave, how we communicate, and how we think.

Even two millennia ago, in Luke's time, one can see that this idea of "what goes around comes around" was already commonplace, certainly among the authors of the Gospels. Mark 4:24 says: "And he said unto them, Take heed what ye hear: with what measure ye mete, it shall be measured to you: and unto you that hear shall more be given." Matthew 7:2 puts it another way: "For with what judgment ye judge, ye shall be judged: and with what measure ye mete, it shall be measured to you again." What is remarkable is that all kinds of fascinating consequences spin out of this perspective.

In a small group, say a village, what we call indirect reciprocity bestows tremendous advantages, by allowing me to benefit from the experience that others in our clan had when dealing with you. ("Ugg has always been fair when it comes to trading tools for food. But Igg can't be trusted.") When dealing with you, I take into account more than just our dealings with each other.

While direct reciprocity relies on your own experience of another person, indirect reciprocity also takes into account the experience of other people. Mathematicians could say that indirect reciprocity is a broader category that includes direct reciprocity, but the two mechanisms are analyzed in quite different ways: to dissect the direct form we need to look at repeated games, as we did in the last chapter. To understand the indirect form we need to recognize the power of reputation.

Exploring the indirect form of reciprocity is important because it is critical for society. Direct reciprocity—"I'll scratch your back and you scratch mine"—operates well within small groups of people, or in villages where there is a tight-knit community where it would be hard to get away with cheating one another. In small societies, indirect reciprocity is also at work, as people create, observe, and report the soap opera of everyday life. But by the time of Christ, Eurasia's middle latitudes were straddled by the Roman Empire, the Parthian Empire, the Kushan Empire of Central Asia and Northern India, and the Han Empire of China and Korea. To extend and thrive, these sprawling societies had to depend on more than just direct reciprocity.

Societies could more easily evolve to become larger, more complex, and interconnected if their citizens depended on economic exchanges that relied on indirect reciprocity. Today, this is central to the way we conduct our affairs and cooperate. With the help of gossip, chat, and banter we are able to gauge the reputation of other people, sizing them up, or marking them down, to decide how to deal with them. This sheds light on both the proliferation of charity and of glossy celebrity gossip magazines.

Thanks to the power of reputation, we think nothing of paying one stranger for a gift and then waiting to receive delivery from another stranger, thanks also to the efforts of various other people whom we have never met and will never meet—from the person who packs our gift to the one who checks our credit rating. In our vast society it is a case of: "I scratch your back and someone else will scratch mine." We all depend on third parties to ensure that those who scratch backs will have theirs scratched eventually.

Under the influence of indirect reciprocity, our society is not only larger than ever but also more intricate. The increasing size of modern communities can now support a greater subdivision of physical and cognitive labor. People can specialize when networks of indirect reciprocity enable a person to establish a reputation for being skilled at a particular job. Thanks to the power of reputation, great collections of mutually dependent people in a society can now sustain individuals who are specialized to an extraordinary degree, so that some of its denizens are able to spend much of their time thinking about how to capture the quintessence of cooperation in mathematical terms while others are paid to think about how to express mathematical terms about cooperation in plain English. It's amazing.

This link between the size of a settlement and the specialization of its inhabitants was recorded in ancient times. Xenophon, an Athenian gentleman soldier, wrote in the fourth century BC that the bigger a settlement was, the more finely divided its labor: "In a small city the same man must make beds and chairs and ploughs and tables, and often build houses as well; and indeed he will be only too glad if he can find enough employers in all trades to keep him. Now it is impossible that a single man working at a dozen crafts can do them all well; but in the great cities, owing to the wide demand for each particular thing, a single craft will suffice for a means of livelihood, and often enough even a single department of that; there are shoe-makers who will only make sandals for men and others only for women. Or one artisan will get his living merely by stitching shoes, another by cutting them out, a third by shaping the upper leathers, and a fourth will do nothing but fit the parts together."

BRAINPOWER AND INDIRECT RECIPROCITY

Indirect reciprocity is not only a mechanism for the evolution of cooperation but also provides the impetus for the evolution of a big brain. To explain why, I should once again emphasize that cooperation means paying a cost for someone to receive a benefit. Thus, in effect, we buy a reputation. For example, it costs you precious time when you come to the aid of a stranger so that you may end up being late for that pressing appointment with your boss. Or if you give a lift to someone whose automobile has broken down, you could end up with a smear of motor oil on your new silk tie. But the point is that this little generous act secures you a reputation, which might be worth a great deal—more than the initial cost—in the long run.

Thanks to the power of reputation, we help others without expecting an immediate return. If, thanks to endless chat and intrigue, the world knows that you are a good, charitable guy, then you boost your chance of being helped by someone else at some future date. The converse is also the case. I am less likely to get my back scratched, in the form of a favor, if it becomes known that I never scratch anybody else's. Indirect reciprocity now means something like "If I scratch your back, my good example will encourage others to do the same and, with luck, someone will scratch mine."

By the same token, our behavior is endlessly molded by the possibility that somebody else might be watching us or might find out what we have done. We are often troubled by the thought of what others may think of our deeds. In this way, our actions have consequences that go far beyond any individual act of charity, or indeed any act of mean-spirited malice. Our behavior is affected by the possibility that somebody else might be watching us. We all behave differently when we know we live in the shadow of the future.

That shadow is cast by our actions because there is always the possibility that others will find out what we have done, whatever the society: it could be the man from the local village gazing down on you from a hill when you helped an old lady; or the woman who was walking by when you carried all those groceries for your wife; or the boy who came to deliver a gift to a neighbor; or the guy who sits at the adjoining desk; or the security guard looking at you through a closed-circuit camera. Each of us also wants our friends, family, parents, and loved ones to know that we are good, helpful people. In coming to the aid of another, or letting another person down, you not only help develop your reputation; you also help perpetuate and bolster the complex and tangled web of indirect reciprocity essential for a large, complex society to run smoothly.

For many people to appreciate your selfless act, and for your reputation to flourish, we need more than language. We need smart and receptive brains. Indirect reciprocity relies on what others think of us. Making a reputation has been shown to engage much of the same reward circuitry in the brain as making money. By being helpful, I obtain the reputation of being a nice, obliging, and considerate person. My behavior toward you, of course, now depends on your reputation and thus what you have done to others: if you have been a cad and a rotter, I am less likely to trust you to deliver. Then again, if we know nothing about someone, we are often willing to give them the benefit of the doubt for the sake of our own reputation.

There is a clear link between this mechanism of cooperation and the evolution of empathy. We need to have a good idea of what is going through the mind of another person in order to understand and appreciate the motivation of a Good Samaritan. "Even though he was rushing home to see his ill mother, he stopped to help that injured man," "If I had been lying there, bleeding at the curbside, I would have been so grateful for the help of a stranger," "I could see she was in pain and felt I had to help," and so on. We require, in the parlance of the psychologists, a "theory of mind," that remarkable capacity that enables us to understand the desires, motivations, and intentions of others. This mind-reading ability allows us to infer another's perspective—whether emotional or intellectual.

One can easily envisage how the mechanism of indirect reciprocity can stimulate the evolution of moral systems. The quotation from Luke at the start of this chapter has a direct corollary, known as the Golden Rule, that transcends all cultures and religions: "Do unto others as you would have them do unto you." The rule pops up in Greek philosophy ("What you wish your neighbors to be to you, you will also be to them," Sextus the Pythagorean), Buddhism ("Putting oneself in the place of another, one should not kill nor cause another to kill"), Christianity and Judaism ("Love your neighbor as yourself"), in the Mahabharata of Hinduism ("One should never do that to another which one regards as injurious to one's own self") in the Farewell Sermon of Muhammad ("Hurt no one so that no one may hurt you") and in Taoism too ("He is kind to the kind; he is also kind to the unkind").

The Golden Rule interlinks several ideas: it binds empathy with the idea of reciprocity along with an ironclad faith in the power of indirect reciprocity—if I am good to another person today, somebody will be good to me in the future. In this way, indirect reciprocity has played a central role in the development of our brains, of our ability to lay down memories, and of our language and moral codes. This remarkably potent ingredient of cooperation is at the heart of what it means to be human.

KAHLENBERG

I first came to appreciate the power of indirect reciprocity on a walk with Karl Sigmund in the summer of 1996. We were hiking around the Kahlenberg, a forested hill north of Vienna. The ridge is blessed with breathtaking views of the great city, being part of the Wienerwald, the Vienna woods. We were negotiating a chain of tree-covered hills northeast of the city, bounded by rivers, including the Danube. There are villages here and there, such as one where Beethoven lived (Nussdorf), and taverns (*Heurige*) are dotted all about, where we could sit down to sip the local wines.

Although this does not sound a likely birthplace for a scientific breakthrough, there is plenty of evidence that the networks of paths that crisscross the forested hills of the Wienerwald are steeped in creative magic. Mahler would walk from the Kahlenberg into the city to conduct opera. Johann Strauss the Younger composed his "Tales from the Vienna Woods" in waltz time. Franz Schubert and Beethoven were also moved by its rolling Arcadian landscapes. A green meadow on a plateau high above the city where the skies open wide is called Himmel (Heaven). There, the young Sigmund Freud managed to convince himself that he had understood the nature of dreams.

During our long meander though heaven, Karl mentioned something that made me stop in my tracks. He suggested that we should extend our work on cooperation to take a close look at *indirect* reciprocity. I had never heard that expression before but so many thoughts gushed into my mind that I found it intoxicating. I told him that I did not want him to explain too many details. I did not want to know what work might have been done before on this subject, so I could follow through my own line of thinking. I knew exactly what he meant and how the perfect clarity of mathematics could bring this idea into sharp focus. I stopped everything else that I was doing. In my mind a rolling landscape of new possibilities for cooperation beckoned.

I fell in love with this work, which I felt would take our research in a new direction. I was almost consumed by the feeling. Amour did seem to be in the air. For one thing, I was reading *The English Patient* ("In love, there are no boundaries"). For another, Karl and I had made a poignant discovery during our walk in the lush greenery of the Wienerwald. We stumbled across a little cemetery, where there was an overgrown grave. The headstone was carved with poems and stories to celebrate the memory of Caroline Traunwieser, apparently the greatest beauty of the Vienna Congress of 1815.

Among the many dedications to Caroline was a tribute from the founder of the Austrian Academy of Sciences, Freiherr von Hammer-Purgstall, an Orientalist scholar who supplied Goethe with Persian poetry. He recounted his first heart-skipping encounter with her in a salon: "Never before and never afterwards in my life was I so overwhelmed by the appearance of beauty." She was adored by all, from poets to officers to the director of the Viennese China Factory (Wiener

Porzellanmanufaktur). Her grave revealed that she died young and that no portrait of her survived. In a strange way, Karl and I felt bereft after reading these moving tributes to her lost beauty. My melancholy was a faint reflection of Caroline's radiant glow that had illuminated Vienna long ago, a testament to her reputation.

FROM REPUTATION TO COOPERATION

The most incomprehensible thing about the universe is that it is comprehensible.

—Albert Einstein

By the time my flash of inspiration came in the Wienerwald, my confidence in my ability to crack problems was growing. Deep in my brain a geyser bubbled and sent a torrent of thoughts skyward. I knew I had to work fast. Nearby was my parents' house, on the northern slope of the Kahlenberg. In my little bedroom, which I have used since I was eight years old, I sat down and began my research on indirect reciprocity.

Usually when you begin a new project you immediately run into difficulties. There's one unforeseen problem, then another. Often there are many. You need time to wrestle with them and, only if you are very lucky, resolve them. Usually you fail. Not this time, however. Everything I attempted worked, and first time too. After three weeks, I had an almost complete story, a mathematical picture of indirect reciprocity and, most important, how it helps cooperation to bloom. I was driven on by the excitement of trying and succeeding at something new. I was proud of the lightning speed with which I managed to knit my intuitions into a mathematical theory.

After three weeks I saw Karl again to discuss my findings. Once again, we met in the forest. This time, the weather was grey, the air damp and raw. We had arranged to meet in a little inn and, once we had sat down together at a wooden table, I presented my results to him. Even though we were friends, I felt nervous as though I were revealing a secret for the first time. Karl liked the approach and immediately saw the implications.

I began with a computer model that described a population of people. In that population, any one encounter involves two people. One of them is offered a choice—whether or not to help the other. When a Good Samaritan does something nice for someone else, this altruistic act confers a benefit on the recipient at a cost to the Samaritan. That could be when you have to sacrifice your time to help another, whether giving a hand to that doddery old lady to help get her safely across a street or taking a moment to point out the nearest car park to a motorist.

If the cost is smaller than the benefit then the act of charity, once returned, leaves both individuals better off. This puts us in very familiar territory. One can think of this setup as a simplified version of the Prisoner's Dilemma that we explored in chapter 0. Cooperation means paying a cost for the other person to receive a benefit. Defection means doing nothing. If you think of one person as the donor and the other as the recipient, then it adds up to half of that problem, a demi Dilemma.

As we saw in the case of the Prisoner's Dilemma, it is rational to defect. But only in a single encounter. If our players see each other again and again, cooperation can emerge because rational players must weigh the benefit of exploiting the other player in the first round against the cost of forfeiting collaboration in future rounds. But, of course, repeated encounters between the same two players leads to direct reciprocity. I now wanted to study the evolution of cooperation in a more general, indirect setting.

I arranged the game so that each player can take part in many rounds, but typically not with the same partner twice. Thus a defector—someone who does nothing to help—could not be held to account by an earlier victim. But defection can be detected nonetheless, as a result of players building up a reputation: a player's reputation score (Karl and I called it "image" in our paper) is 0 at birth and rises whenever that player helps others. Equally, it falls whenever the player withholds help. This is an important ingredient of the game. It meant that we did not divide our players up into goodies or baddies. Instead, we graded the image of each player so it could be more nuanced and change as the game evolved.

There were also unconditional cooperators and unbending defectors. I built one more feature into the model, to add a touch more realism.

Just as only a certain group of people are privy to gossip, so the outcome of any given encounter between players is only revealed to a subset of people in the population. As a result of this, different people hold different views about the reputation of the same person.

Karl and I found that if the cost-to-benefit ratio of cooperation is sufficiently low, and the amount of information about the co-player's past sufficiently high, cooperation based on discrimination—favoring good reputations—can emerge. Instead of relying exclusively on my direct experience with someone (as is the case for direct reciprocity), I can now also benefit from the experience of others. Now my behavior toward you not only depends on what you have done to me, but also on what you have done to others.

The bottom line of these evolving populations was the following: if there is enough transfer of information about who did what to whom from person to person, then natural selection favors strategies that base their decision to cooperate (or defect) on the reputation of the recipient. If good reputations spread quickly enough, they can increase the chances of cooperation taking hold in a society. And, as one would expect, Bad Samaritans with a poor reputation receive less help.

We were not the first to argue that reputation is possibly an important factor for altruistic behavior. A form of the concept had been articulated—in a verbal outline, rather than mathematically—in a book by Richard Alexander at the University of Michigan, an expert on crickets, katydids, and cicadas. It was Alexander who, in *The Biology of Moral Systems* (1987), had first coined the term "indirect reciprocity." Alexander had posed tough questions such as, What is moral? or How do we start to crystalize our ideas about what is good and bad? He argued that the answer lay in reputation. We are always scanning the impressions left by others and are more likely to give to somebody who has a good reputation, someone who has in her or his past given help to others—not necessarily to me, though, but just to somebody. Indirect reciprocity "involves reputation and status, and results in everyone in the group continually being assessed and reassessed." This, he said, plays an essential role in human societies.

The idea appears in the work of economist and philosopher Robert Sugden of the University of East Anglia. He put forward the concept of "standing" in *The Economics of Rights, Co-Operation, and Welfare* (1986). The idea goes like this: If you defect against somebody who is in good standing, you move into bad standing. But if you defect against somebody who is in bad standing, you remain in good standing. A mathematical depiction of social norms had also been explored by the Japanese economist Michihiro Kandori. The fact that our new theory of indirect reciprocity had reputable forerunners gave it further credibility.

Karl also had plenty of anecdotes to underline why indirect reciprocity is just as relevant to everyday life as the direct version of reciprocity. He pointed out that the Rothschild family had protected the investments of their English clients during the Napoleonic Wars. They were under intense pressure to give them up, but they kept the interests of their English clients at heart. Afterward, of course, the Rothschild family became extraordinarily rich. Their fortune was a direct result of the power of indirect reciprocity: because they had behaved impeccably, everybody now knew that they could be trusted.

Then there was the story of the American baseball player Yogi Berra, who was famous for his pithy comments and witticisms known as Yogi-isms. One of them was a perfect summary of indirect reciprocity: "Always go to other people's funerals, otherwise they won't come to yours." Berra was counting on the fact that his acts of kindness would not be returned by the recipients, but by third parties who were moved by his public mourning.

The idea is also wonderfully summarized in musical form by Tom Lehrer, the American singer-songwriter, satirist, pianist, and mathematician. In "Be Prepared," Lehrer's spiky salute to the Boy Scouts, he sings: "Be careful not to do / Your good deeds when there's no one watching you." German speakers even have a saying with the same gist, Milinski notes: *Tue Gutes und rede darüber.* ("Do good and talk about it.") The converse was of course true too. This may all sound obvious. But without a mathematical model, we would have no quantitative understanding of how this mechanism really works. Nor would we be able to reveal the many subtleties of indirect reciprocity. The time was ripe to pin down the idea.

When, for example, Karl and I made the simulation more realistic and allowed for mutations, or mistakes in an evolving population of players, then we saw cooperation and defection wax and wane over time, as those with a good reputation are actually undermined by indiscriminate altruists who help anyone, no matter how well or badly the latter have behaved in the past. Then, free riders—unconditional defectors—invade until discriminating cooperators cycle back in. Given my earlier work on the Prisoner's Dilemma, I was not surprised by this. But anyone unfamiliar with the field would have found it striking how the degree of cooperation always rides a seesaw of cycles.

Importantly, we found that natural selection favored strategies, called Discriminating strategies, that pay attention to the reputations of others. These strategies prefer to interact with people who have a good reputation. Thus natural selection (acting in the framework of indirect reciprocity) promotes social intelligence: observe others, learn about them, understand who did what to whom and why.

Karl and I also made an intriguing discovery that underlined how, when people act on their convictions, it can come at a cost. Refusing help to a free rider or other defector lowers the score of discriminating players so that, even though they may have acted for good reason, they might come over as Bad Samaritans. A colleague at work lets you down badly and you snap at her. From the perspective of fellow workers in a calm open plan office, your angry outburst makes you appear out of control. Or you may decide not to help a tramp because he whispered an insult at you. To an onlooker on the other side of the road, however, it looks like you have turned your back on a poor and deserving vagrant. This also diminishes your likelihood of being helped in turn.

The bottom line of our theory was that an act of altruism will only evolve when the shadow of the future—that is, the expectation of coming gains—exceeds the cost. This idea could in turn be summed up by a simple mathematical relationship. The evolution (emergence) of cooperation can occur if the cost-to-benefit ratio is exceeded by the probability of knowing someone's reputation, if you like. Karl and I submitted our work to the prestigious journal *Nature*. The paper was accepted without much bother and was published in 1998. In its wake came many more papers on the subject of indirect reciprocity, including experimental confirmation.

In this way, our walk on the Kahlenberg had turned out to be a eureka moment, one of the most romantic and best-known feelings described in accounts of research. The vanishingly unlikely part of a eureka story is not the pounding heartbeat that comes with a novel insight but the awareness that you really did have a Big Idea, one that had an impact. And there's the rub. This tends to creep up rather slowly in science. Karl and I were lucky because sometimes the full significance of a eureka moment will only emerge much later. In fact it can often take years for an idea to become concrete. Sometimes longer than a lifetime. I was once moved by a biographer's words about the Austrian composer Franz Schubert: how "a later world would give him his due, slowly though it came to him at first."

THE EVIDENCE

There's a telling joke among scientists that every new theory has to pass through three phases of "acceptance": first, it is completely ignored; second, it is obviously wrong; and third, it is obviously right, but everyone knew that anyway. Karl and I were fortunate. We did not become the punch line of this old joke—at least not this time.

A couple of years after our walk, we found ourselves writing a comment for the journal *Science* on a clever piece of experimental research that had provided backing for our *Nature* paper on indirect reciprocity. Working at the University of Bern in Switzerland, Claus Wedekind and Manfred Milinski had started out with seventy-nine first-year students who were blissfully unaware of concepts such as reciprocal altruism and invited them to take part in a game in which they had the option to donate money to other individuals in the group.

The game consisted of encounters between pairs of students who were connected by a computer network. One student was the "donor," the other the "recipient." If the donor paid one Swiss franc out of his account, the recipient would receive four. Thus the cost for the donor was 1 SFr and the benefit for the recipient was 4 SFr. As ever, for productive cooperation the benefit must exceed the cost. Alternatively the donor could decide to pay nothing and, of course, the recipient would not receive a bean. When making his decision about whether to give or to hang on to his money, the donor was informed about what the recipient donated in previous rounds. For example, a donor could learn that her recipient was stingy and never gave a thing, or was relatively generous and gave two out of three times. In order to exclude the effects of direct reciprocity, the experiment was arranged in such a way that the same two students did not meet again.

The outcome of the experiment was convincing. Wedekind and Milinski found that even when there is no chance of direct reciprocity players are generous to each other provided that they have an opportunity to keep tally of the actions of their fellow player. We cooperate more with those who have a good reputation. As a result, people who started off by being generous ended up with a high payoff. We like to give to those who have given to others. Give and you shall receive!

THE MORAL SPECTRUM

Let's examine one subtlety of my computer simulation of indirect reciprocity. If you see a Bad Samaritan and refuse to help him, you yourself could end up looking like another Bad Samaritan who in turn would be rejected by others (even though you had a very good reason to be a Bad Samaritan). A smarter rule should distinguish between justified and unjustified defections and should therefore take into account the reputation of the receiver too: someone withholding help from a "bad" player should not damage his own reputation as a result.

One way to extend the work Karl and I had done was to study the effects of these more sophisticated rules. To make the problem tractable, it helps to assume that there are only two kinds of reputation: good and bad. In this world of binary moral judgments there are four ways of assessing donors in terms of "first-order assessment": always consider them as good, always consider them as bad, consider them as bad if they give and good otherwise, or consider them as good if they give and bad otherwise. Only the last option can lead to cooperation based on good reputation.

Second-order assessment rules take into account the reputation of the receiver too, so we are now able to consider the wider circumstances; as mentioned already, it can be deemed good to refuse help to a bad person. There are sixteen of these second-order rules. There are also third-order rules, which depend additionally on the score of the donor (after all, a person with a poor reputation might try to "buy" a good one by being more generous to those with good reputations). And so on. In all, there are 256 third-order rules.

Once we have assessed the players to the first, second, or whatever order we decide, we then have to work out what to do. Do we give help, or do we walk on by? This is decided by a so-called action rule. The action rule depends on the recipient's score and on one's own (there are four possible combinations of the two scores and thus a total of sixteen action rules). For example, you may decide to help if the recipient's score is good or your own score is bad. You might reason that by doing this you might increase your own score and therefore increase the chance of receiving help in the future.

A strategy is the combination of an action rule and an assessment rule. Given the above, we obtain 16 times 256, which is 4,096 strategies. That is a lot. Nonetheless, this universe of strategic possibilities has been explored in a remarkable study at Kyushu University in Fukuoka that was the basis of the doctoral thesis of a brilliant theoretician, Hisashi Ohtsuki, who will also feature in later chapters.

Ohtsuki's adviser was the formidable Japanese mathematical biologist Yoh Iwasa. During my first visit to Japan, almost every person I met had introduced himself or herself as being a student of Iwasa. I became curious about this beacon of inspiration. I wanted to meet the professor who was "the number one in Japan." Yoh himself always jokes that, while most Japanese names mean "the great" or "the brilliant," his actually means only "the mediocre." But again he's being modest. In fact, his name signifies "the golden mean," the most desirable position of perfect balance. He and his students are a major force in theoretical biology.

In search of the perfect balance of qualities to ensure that indirect reciprocity helps us all to get along, Ohtsuki and Iwasa analyzed all 4,096 possible strategies and proved that only eight of them are evolutionarily stable and can lead to cooperation. The leading strategies share some characteristics: cooperation with a good person is regarded as good, while defection against a good person is regarded as bad. In other words, as one would suspect, they differentiate between justifiable and nonjustifiable defection, that is, they discriminate against Bad Samaritans. If a good donor meets a bad recipient, the donor must defect and this action does not reduce his reputation. Instead, it is seen as a "justified sanction." Similar problems were also explored by Karl Sigmund and Hannelore Brandt in Vienna.

There are still many open questions regarding indirect reciprocity. To illustrate one of them, consider the following scenario, one that seems relevant to today's celebrity-obsessed culture, in which people are now famous for being famous. If I help someone with the sole purpose of increasing my reputation, what then? We place a major emphasis on not only monitoring what other people do, but also on understanding their motives. If they do something flamboyant and showy without any real concern for the well-being of others, then something is wrong. I find this unanswered question fascinating and it brings us back to the Gandhi quotation at the start of this chapter.

STILL WALKING IN THE WIENERWALD

Who steals my purse steals trash; 'tis something, nothing;
'Twas mine, 'tis his, and has been slave to thousands:
But he that filches from me my good name
Robs me of that which not enriches him
And makes me poor indeed.

—Shakespeare, *Othello*

During our hike in the Vienna woods, Karl and I had discovered another mechanism for the evolution of cooperation, one that relied on reputation. In direct reciprocity, all I can do is to learn from repeated encounters with the same person. As a result, my behavior depends on what you have done to me. But for indirect reciprocity, there are repeated interactions in a group. Now my behavior toward you also depends on what you have done to others.

This idea is now abundant in e-commerce, where there are many applications. The web abounds with ways to score the behavior of people, for example. Even when you encounter a stranger, you can benefit from someone else's experience with him or her. Now, when buying a camera online, you consider the seller's reputation as closely as the price. Subscribers to eBay auctions are asked to state, after every transaction, whether they were satisfied with their partner or not. Their partner's score can accordingly increase or decrease by one point. The ratings of eBay members, accumulated over twelve months, are also public knowledge. This crude form of assessment seems to suffice for the purpose of reputation building and seems to be reasonable proof against being manipulated.

As a result, I can benefit from the experience of others when dealing with you by paying close attention to your reputation. If you have been unreliable, then I will be wary. However, if you have been generous, I am more likely to work with you. In this way indirect reciprocity is a powerful promoter of cooperation. David Haig, an evolutionary biologist at Harvard, sums this up elegantly: "For direct reciprocity you need a face. For indirect reciprocity you need a name."

And for a name you need language, a convenient label to distinguish one person from another in the great game of life. Thus for indirect reciprocity to work, we need a way to communicate with each other, to discuss our hopes and our fears, and to learn from the experiences of others. I believe that the demand for social cooperation via indirect reciprocity has, more than anything else, propelled the evolution of human language. And to possess a faculty as complex as human language, you need an impressive brain. My stroll with Karl had taken us a long way from direct reciprocity to reveal a vast new panorama of cooperation.

CHAPTER 3

Spatial Games—
Chessboard of Life

*The chess-board is the world, the pieces are the phenomena of the universe,
the rules of the game are what we call the laws of Nature. The player on
the other side is hidden from us.*

—Thomas Huxley

*If people do not believe that mathematics is simple, it is only because they
do not realize how complicated life is.*

—John von Neumann

Where there's life, there are lumps, clumps, and colonies. Bacteria grow
in films. Slime molds aggregate in three-dimensional shapes, similar to
Mexican hats. Bisons gather in herds. Ants work in colonies. Apes form
troops. There are sleuths of bears, murders of crows, pods of whales, and
gaggles of geese. And, of course, populations of people form structures
too. We are organized in villages, towns, and cities. We gather in the
workplace, at schools, and in theaters and pubs. We have mobs and
crowds and posses and throngs.

When I started to work on ways reciprocity can solve the Prisoner's Dilemma, I wondered if population structure could offer another
way to solve the Dilemma. Remember that the calculations of the past

two chapters were based on a simple assumption: the players in the Dilemma were in a well-mixed population, where every player has an equal chance of meeting every other player. In these uniform populations we found that defectors always outcompete cooperators. But, of course, all real populations have some structure. What difference does this make? Could population structure affect the outcome of the simple Prisoner's Dilemma? Could there be structures that make cooperators triumph over defectors?

The very fact that people decide to live in the same patch, rather than dotted around at random, has to do with cooperation. Why? Some believe, for example, that the first communities evolved as a result of success in agriculture: surplus food from farming enabled people to settle down and specialize, from butcher to baker to candlestick maker. Others link it to ancient belief systems and religions. At Göbekli Tepe in Turkey ("Hill with a potbelly"), for example, is a sanctuary with limestone pillars that were carved and erected by hunter-gatherers more than eleven thousand years ago. That remarkable discovery would suggest that temples planted the seeds of cities.

Perhaps cities were born of the struggle for existence: in his book *Whole Earth Discipline,* Stewart Brand suggests that the very first urban invention was the defendable wall, followed by rectangular buildings that could pack people efficiently inside that wall. The Cambridge archaeologist Colin Renfrew argues that the first communities came with the birth of the modern mind. That is when the effects of new intellectual software kicked in, allowing our ancestors to work together in a more settled way. However, I would discover from my computer simulations that you do not need any brainpower at all to benefit from forming a huddle.

THE GOD GAME

One can trace the efforts to link life and geography—in the guise of what are called spatial automata—to a study by the great John von Neumann, who believed that biological organisms could be thought of as information-processing systems. The fact that it is now possible to write and synthesize the genetic code of a living thing in the laboratory, like a glorified computer program, shows how right he was. He puzzled over the difference between the trivial kind of reproduction that enables crystals to grow in a test tube, for example, and the clever kind that enables creatures to breed. To explore the difference, von Neumann wanted to design a machine that was complex enough to reproduce itself.

To this end, he devised a "self-reproducing automaton," a robot that was afloat on a sea of its own components, just as living things on Earth thrive and abound among the chemical building blocks of life. He built on the work of Alan Turing, the English mathematician who had laid the logical foundations of computing with the idea of the "universal Turing machine," which offered a splendid abstract device to explore the theoretical limits of mathematics.

Von Neumann showed that there also exists a universal automaton, an abstract simulation of a physical universal assembler. Logical errors within the automaton could be viewed as mutations, allowing the possibility of more complex varieties of automaton to emerge; in an environment with finite resources, selection pressure—survival of the fittest—would lead to Darwinian evolution. But there was no mathematically rigorous way to analyze his creation, let alone the means to build one.

The mathematician Stanislaw Ulam gave von Neumann advice on how to make his self-reproducing machine simpler. He put forward a way that Neumann's "machines" could be built of pure logic. Ulam suggested replacing the floating automaton with what he christened "tessellation robots." This quaint term refers to the growth of crystals, which occurs by the buildup of unit blocks, or "tessera." Today Ulam's tessellation robot is called a "cellular automaton" and consists of an abstract array of cells programmed to execute rules en masse. This collection of cells—like squares on a chessboard—carries out computations in unison and can be viewed as a kind of organism, running on pure logic, though the cells in question have little to do with the real thing.

Each cell in the array is in a particular "state" at a given time. A state might be a certain color, say red, green, or blue, a numeric value, or simply on or off. Time is not continuous in the automaton, but discrete, so that it advances with each tick of a clock. With each tick, simple rules determine how any cell changes state from one instant to the next. These rules depend not just on the state of that particular cell but also on the states of its neighbors—a predetermined "look-up" table based on rules decides what the cell has to do next. So, in a black-and-white automaton, it could be that if all neighboring squares are black, the central square is made white. Despite the apparent simplicity of this setup, it turned out that anything achieved in von Neumann's original automaton could be aped by Ulam's cellular system.

After von Neumann's death, the torch for cellular automata passed to others, notably to John Holland, professor of Psychology and professor of Electrical Engineering and Computer Science at the University of Michigan, Ann Arbor. In 1960, Holland outlined an "iterative circuit computer" related to cellular automata that could mimic genetic processes. Despite the dry title, this research captured the public imagination. As one newspaper put it, "He's the man who taught computers how to have sex." This logical view of life now began to take hold and multiply in the minds of researchers in other laboratories. One of the best known is the British mathematician John Conway.

For several months in the late 1960s, the mathematics department at Cambridge University was taken over by Conway's efforts to find a hypothetical machine that could build copies of itself. From a small table an assortment of poker chips, foreign coins, cowrie shells, and whatever else came to hand was used to mark out the "living" squares in patterns that spread across the floor of the common room as the rules were enacted during each tea break. Even then, Conway used a computer to study particularly long-lived populations.

In 1970 he unveiled his Game of Life. Now Conway had selected the rules of his automata with great care to strike a delicate balance between two extremes: the many patterns that grow quickly without limit and those that fade away rapidly. The evocative name reflected Conway's fascination with how his combination of a few rules could produce global patterns that would expand, morph shape, or die out unpredictably.

Some people call it the "God Game" in recognition of how players have a toy universe at their beck and call. Taking part is easy, requiring no miracles, religious followings, or holy books. Imagine a checkerboard with counters in a few of the squares. Then follow these simple rules. If an empty square has three occupied neighbors (this includes the diagonal as well as adjacent sites), it comes "alive," nurtured by its neighbors. If a square has two occupied neighbors, then it remains unchanged. Finally, if an occupied square has any other number of occupied neighbors, then it loses its counter—to anthropomorphize, the cell dies pining for neighborly love and the chance to cooperate.

Conway conjectured that no initially finite population could grow in number without limit and offered fifty dollars for the first proof or disproof. The prize was won in November 1970 by a group in the Artificial Intelligence Project at the Massachusetts Institute of Technology, which discovered a "glider gun," a pattern that every thirty moves ejects a glider, a moving pattern consisting of five counters. Since each glider added five more counters, the population could grow without limit. Intersecting gliders were found to produce fantastic results, giving birth to strange patterns that in turn spawned still more gliders. Sometimes the collisions expanded to digest all guns. In other cases, the collision mass destroyed guns by "shooting back."

The patterns produced can be complex; indeed the cellular automaton can be shown to be equivalent to a universal Turing machine, so the Game of Life is theoretically as powerful as any computer. It does not take much imagination to see that cellular automata provide a powerful tool to study the patterns of nature. Stephen Wolfram—who wrote his first paper at the age of sixteen and went up to study at the University of Oxford at the age of seventeen, and who devised a highly successful software application called Mathematica as well as a new kind of search engine (Wolfram Alpha)—turns this idea on its head in his book *A New Kind of Science.* He claims that complexity and randomness in life, the universe, and everything are the outcome of cellular automata. Everything can be thought of as a spatial game.

SPATIAL GAMES

Around the start of the 1990s, while still at Oxford, I began to think about the effects of space on my research. Until then, the players in my computer games had grazed in well-mixed populations where all the players bump into each other at random. But many origins-of-life scenarios envisage our ultimate beginning taking place in a precise spot, where an accident of inorganic chemistry crossed the threshold between death and life to create organic chemistry.

One of the best-known examples can be found in a letter sent by Charles Darwin in 1871 to his botanist friend Joseph Hooker: "But if (and oh! what a big if!) we could conceive in some warm little pond, with all sorts of ammonia and phosphoric salts, light, heat, electricity, etc., present, that a protein compound was chemically formed ready to undergo still more complex changes, at the present day such matter would be instantly devoured or absorbed, which would not have been the case before living creatures were formed."

Here Darwin is talking about a scenario that I will return to in detail in chapter 6, where I consider the struggle between what I call prelife and life itself. In Oxford, I approached the idea about life's origins from another angle. I began to puzzle over how to build geography into game theory. What would happen, I wondered, if you played out the Prisoner's Dilemma, or indeed any other game, among players who are dotted about on a landscape? I began to think of Darwin's "protein compounds"—or whatever critical molecules paved the way for life—as players in one such spatial game. If they cooperated, life could emerge.

In outline, it is easy to see how a spatial game could work along the same lines as the cellular automata. The game players are arranged on a chessboardlike array (it can be in three dimensions, of course, or even more). During each round, the player on a given square plays the game with its neighbors. After this, each square is occupied by its original owner or by one of the eight neighbors, depending on who won that round—in other words, who got the biggest payoff.

One can also appreciate how simply thinking about how molecules, cells, or whatever are arranged in space—and thus relative to each other—marks an important change from usual work on the origins of life. Many origins scenarios envisage chemical reactions taking place in a well-mixed medium, whether in the turbulent, superhot mineral rich waters of a deep hydrothermal vent in the ocean or the middle of Darwin's "warm little pond," rather than in the mud around its edge or floating in scum. This is the primordial soup, and generations of theorists have tried to figure out its composition and assumed that it is homogeneous—that it would be the same wherever you chose to dip your spoon.

But there are origins theories where spatial organization is important. Some people suggest that early steps in genesis could have occurred in spots on the surface of rocks, or even between layers of clay, where molecules can concentrate and join together into long chains. A consequence of this is that there could be clumps of different composition, reflecting how different chemistry can occur on surfaces or between clay layers. The influential evolutionary biologist John Maynard Smith coined the culinary term "primordial pizza" to capture the spottiness of this idea. When I started my work in Oxford, I became a pizzaiolo, a primordial pizza chef.

I could draw on earlier work for insights into the effects of space and geography. In the late 1980s, my mentor Bob May had done studies with Michael Hassell of Imperial College, Silwood Park, on wasps that lay their eggs in or on insects which revealed there was a patchiness in the distribution of predator and prey. The two felt they had both made critical contributions to spatial ecology and, typical of Bob, he proposed that they settle the order of authorship on a joint paper with a twenty-five-game series of croquet. (Bob ruefully remarked after their tournament: "The paper, I regret to tell you, is Hassell and May.") They were trying to explore the effects of geography on chaos theory, which shows how the apparently random waxing and waning of populations is predictable in the short term.

For my research on the Prisoner's Dilemma, I decided to use a simple approach. I would simulate a patchy ecosystem by playing the game across the squares on a chessboard. I programmed a computer to find out what happened in what I called the "spatial Prisoner's Dilemma," using a four-color code to help reveal what was going on. There, the fate of a single square depends on its own strategy, those of the eight neighbors, and the strategies of their neighbors too. That is, its fate is sealed by twenty-five squares in all, a five-by-five array. The fate of each square after each round of spatial games was announced to the world by its color.

The results were immediately rich, complicated, and compelling. This was, in itself, fascinating because I had made the games as simple as possible. There was no Tit for Tat, no reputation, and no conditional behavior. I only allowed for pure unconditional (stupid) cooperators and pure incorrigible (rotten) defectors. My move to strip the universe of possibilities down to the bare minimum of goodies and baddies was quite deliberate.

If you allow direct or indirect reciprocity, then we already know that cooperation and complexity can evolve. But if you exclude them then you are stuck with the simple one-shot (nonrepeated) Prisoner's Dilemma, which never leads to cooperation in a well-mixed population. There was no chance that cooperation could possibly take hold in this classical setting, a population where every player is equally likely to meet with everybody else. By boiling the model down to its bare minimum, I could investigate the undiluted, uncompromised effect of space on the Prisoner's Dilemma.

I began to explore this brave new world by simply varying the starting configurations and parameter values. Given the computer mantra "garbage in, garbage out" one might have expected "simplicity in, simplicity out." Given what I knew about well-mixed populations, I did not expect much to happen. Yet before my eyes a complex variety of patterns blossomed. Cooperators and defectors could coexist with each other. Some patterns were static and others displayed oscillations, so they underwent cycles of boom and bust.

One day, when cycling home, I realized that a particular setup of my primordial pizza—a squarelike cluster of defectors adrift in a sea of cooperators—could do something special. It should grow on its four corners but shrink on its four edges. If this was really the case, I thought, then something very complicated might well start to grow in my computer. As soon as I got home, I programmed the new scenario into my machine and watched what happened.

The result was astounding. The most amazing patterns unfolded. These evolutionary games generated irregular or regularly shifting mosaics, where strategies of cooperation and defection coexisted in an endlessly milling chaos. From a single defector sprouts a fantastic kaleidoscope of gorgeous patterns, suggestive of lace doilies and stained glass windows. I thought to myself that I was the first person ever to see this vital, beautiful pattern. Indeed, because it was ever-changing— dynamic—its turbulent patterns seemed to capture the shifting essence of life itself. I was happy. I was amazed. I wanted to tell someone about it. But there was no one around.

Ursula and I lived in a small apartment at Wolfson College, an academic nirvana that seemed to be full of young, attractive researchers who were passionate about their work. Our apartment overlooked a tranquil punt harbor and a bridge over the Cherwell River, which runs into the Thames. The apartment had only two rooms and both were rather small. In the bedroom, between the bed and the window, I had wedged a square desk. This compact arrangement proved very convenient. Whenever my work was done, I could immediately slump on my bed in relief.

I was fortunate that, back at the zoology building, I could plot and capture these living patterns in Technicolor detail with the help of a computer owned by the evolutionary biologist Bill Hamilton. The machine was a material sign of his high status in Oxford's Zoology Department and it came with a great (for that time) graphics package wired up to what was then extremely advanced hardware: a color printer. Bill was very kind to me, allowing me to tip tap away on his computer in his cluttered office.

I would sit in front of the computer, adjacent to the prized printer, while Bill worked away at his desk behind me. From time to time, we both looked on as these strange patterns bloomed and withered, then sprouted and subsided once again. Bill realized there was something new in the making. We even discussed what would be the best color code to highlight what was going on in the computer: blue for cooperators because it is the color of heaven; red for defectors, because it is the color of the other place. I selected green to be a cooperator that had previously been a defector, and yellow vice versa. In this way, blue and red depicted static cells and green and yellow showed flux and change.

In my games I saw reflections of the extraordinary complexity seen in Conway's Game of Life arise quite naturally. In one game that ran on my computer, a "walker" emerged which consisted of an L-shaped structure of ten cooperators that could stroll without fear through a red sea of defectors. Two walkers could explode into a "big bang" of cooperation if they collided—where splodges of cooperative blue were ringed by green—as defectors were won over to a more obliging way of life.

I found that if I started out with symmetrical patterns, I ended up with ever-changing arrangements of squares that had fractal geometry, where the same kind of structure can be seen on all scales (think clouds and cauliflowers) so that the patterns look the same, even if you magnify them. One could seed such patterns by allowing a single defector to invade a world of cooperators—the defectors gained on corners but lost on the edges. Because the patterns were ever-changing, Bob and I called the never-ending movie a dynamic fractal, a Persian carpet of a design that combined the order of symmetry with chaos. These clusters of cooperators and defectors could keep on growing. Yet, though the actual pattern was fluid and changing all the time, the relative abundance of cooperators always fluctuated around the same level, a mysterious 31.78 percent.

I created a simple mathematical model that tried to explain this observation. I told Bob all about it; and the very next day, as he climbed out of bed, the solution occurred to him. In his brainwave, Bob realized that it required some approximations and a little calculus, which he could do in his head. Finally, it boiled down to a simple integral (which is typically used to figure out the area under a curve drawn on a graph). This, too, he worked out within seconds. All he needed was the value of the natural logarithm of 2. Naturally enough, Bob remembered that the answer is 0.69.

A few moments later and he had the result: 31.78 percent. That was a great moment. In this milling universe of unpredictability and chaos there was one fixed point. The average frequency of cooperators was 31.78 percent. Our calculation had explained why this number holds true for symmetric patterns. However, to this day it remains a puzzle why this magic percentage rules lopsided patterns too, those that form when the combinations of parameters produce the most fluid and interesting behaviors.

Our results were published in 1992, in the journal *Nature*. I can still see the look of joy on Bob's face when our paper was accepted. To celebrate, I went to a little copy shop near New College and had T-shirts printed with the evocative patterns. I foolishly imagined that one day an industry of computer-generated art would be based on the machinations of my primordial pizza program. The entire world would end up crisscrossed with advancing waves of defectors and cooperators. Prestigious galleries such as the Museum of Modern Art in New York would have art inspired by this eternal struggle between light and darkness. Alas, almost two decades later, this particular field of computer-generated art has yet to realize its potential. There is one small consolation. Linux uses one of my patterns as a screen saver.

Although my overblown artistic ambitions were thwarted, there were several interesting conclusions to be drawn. First, once you move from a well-mixed population of defectors and cooperators (a soup) to a heterogeneous population with ghettoes of cooperators and defectors (a pizza) the trajectories of evolution can be very different and cooperation is able to emerge and to thrive, even in the absence of complicated strategies. In other words, in a world of structure, cooperation can emerge without the need for clever thinking—indeed without the need for a brain at all.

In this way, my work told me something profound about the origins of life. The patterns provided evidence that by introducing geography into the Prisoner's Dilemma, cooperators and defectors can exist side by side. In nature, this means that exploiters and exploited, cheats and the honest, abusers and abused can coexist, even without the guidance of a strategy. There is no overall winner in these spatial games of life; rather there is a dynamic interplay of different types.

The simulations represent good news for genesis, which now looks much easier. At the origins of life, it would be much more efficient to plant the seed of cooperation with the help of a "pizza," surface, or structure of some kind, than in the middle of a well-mixed primordial soup. In fact the same applies more generally. This is the take-home message of my work. Clusters of cooperators can prevail, even if besieged by defectors. This is the third mechanism for the evolution of cooperation.

To show this third mechanism at work, I studied a version of the game that was more elaborate. This time, my spatial games had three types of cells: cooperators, defectors, and empty. Here the idea is that only cooperators can invade empty territory—defectors alone die out. This setup gives cooperators a great advantage. One finds that if they get eaten up by defectors then there is a kind of scorched earth effect. The defectors will also die out. When these games play out on the computer, one observes waves: cooperators are chased by defectors, followed by empty squares, which are then repopulated by cooperators.

There are clear implications of this work for more recent stages of evolution, even for quotidian goings-on. Everyday life abounds with spatial cooperation: we are more inclined to be friendly with neighbors. Rather than ask just anyone for a cup of sugar or a carton of milk, it is more natural—and easier—to approach a neighbor. You may have them around for a drink, even if you have nothing else in common other than you live in the same street. In the Vienna of my childhood, shop owners would give discounts to people who worked in nearby stores. This is rich territory for cooperation. In chapters 12 and 13, I will show how I extended my work on spatial cooperation to model the tangled interactions of everyday life, from networks of like-minded people to the effects of friendship and life in the smart set.

CHAPTER 4

Group Selection—
Tribal Wars

I just saw someone who needed help. I did what I felt was right.
—Wesley Autrey, variously known as
the Subway Samaritan, Subway Superman,
The Hero of Harlem, and the Subway Hero.

Mr. Autrey's instinctive and unselfish act saved our son's life. There are no words to properly express our gratitude.
—Larry Hollopeter, father of Cameron

When Wesley Autrey risked his own life to save that of a stranger, he became the epitome of an everyday hero. His good deed made headlines in newspapers and television bulletins around the world. The New York construction worker and Navy veteran was feted by a president and praised by paupers for his inspirational act of selflessness. He showed us all that in the game of life, surviving is as dependent upon the goodwill of others as it is upon personal qualities such as drive, excellence, and so on. He became a living testament to the power of cooperation.

His story of heroism is well known but worth retelling. On January 2, 2007, Autrey was waiting with his two young daughters for a train at the 137th Street and Broadway station in Manhattan when he saw Cameron

Hollopeter undergoing a seizure. Autrey borrowed a pen and used it to keep the jaw of the student open. Moments later, Hollopeter came to and stood up, only to fall down onto the tracks. Despite the lights and rumble that signaled the approach of a southbound train, Autrey jumped down to help the fallen student as his daughters looked on, aghast.

On the tracks, Autrey realized there was not enough time to drag Hollopeter away. Instead, he threw himself over the student to pin him down. The operator of the train applied the brakes, but two cars still passed over the prone figures, close enough to leave a smear of grease on Autrey's cap. Within a few weeks, Autrey had received a flood of gifts, been honored by Mayor Michael Bloomberg of New York City, and showered with praise for his heroic role in the "Miracle Under 137th Street."

Although few of us would feel brave enough to attempt the same selfless act, most of us would at least want to do something to help a stranger in distress. For most of us, it can almost feel like a reflex that does not require any conscious decision making: if we see someone in peril, we become anxious and immediately want to help. This instinct can be so strong that a person will sacrifice his own life to save his comrades. We all have this instinct, even if we don't all act on it. Somehow empathy for the group manipulates individuals, overwhelming their sense of self-interest so they act on behalf of the greater good.

In the context of evolution and natural selection, we have to ask why is this the case. Once again, we have all the ingredients of the Prisoner's Dilemma. We have someone offering help at a cost to himself—life and limb in the case of Autrey—who is a cooperator. What is it that makes us want to hold out a helping hand, rather than hold back and defect? This question has been asked many times before. In fact, this riddle troubled Darwin himself. He put it as follows:

> It is extremely doubtful whether the offspring of the more sympathetic and benevolent parents, or of those which were the most faithful to their comrades, would be reared in greater number than the children of selfish and treacherous parents of the same tribe. He who was ready to sacrifice his life, as many a savage has been, rather than betray his comrades, would often leave no offspring to inherit his noble nature. The

bravest men, who were always willing to come to the front in war, and who freely risked their lives for others, would on an average perish in larger number than other men.

The solution to this mystery could have something to do with indirect reciprocity. (In fact everything about human life has something to do with indirect reciprocity!) A daring and helpful person could get a great reputation among his or her peers, accumulating gifts and the wide-eyed admiration of the opposite sex. But this begs another question: Why does the group react this way? Why do the group's social norms value this kind of extraordinary cooperative behavior? Most likely the social norms of indirect reciprocity require another mechanism of cooperation. Groups with meaningful social norms outcompete other groups. In this way, indirect reciprocity can cooperate with group selection to shape humanity.

Darwin himself believed in this, the fourth mechanism of cooperation. In *The Descent of Man*, published in 1871, he suggested that individuals act for the good of their group. Now the group can thrive at their individual expense:

There can be no doubt that a tribe including many members who . . . were always ready to give aid to each other and to sacrifice themselves for the common good, would be victorious over most other tribes; and this would be natural selection.

In this way, natural selection could encourage cooperative behavior so it would improve the reproductive potential of the group.

Even though this idea is persuasive, many modern biologists came to mock and ridicule the suggestion that groups of organisms might gain a survival advantage over other groups because they shared some beneficial trait. Not so long ago what became known as "group selection" was denounced as a heresy by many evolutionary biologists. The idea was dismissed, trampled, and then swept under the carpet. Critics have said the "group selection fallacy" arises in "amateur misinterpretations of Darwinism" and sometimes even among "professional biologists who should know better." Today, however, such dogmatic attitudes are softening.

With the clear insights that arise from a mathematical understanding of evolution, it is easy to see how natural selection can act both on individuals and on groups of individuals. It is even possible for selection to occur between groups of groups. The simplest situation, in which selection acts on individuals and on groups, can be thought of as selection at two levels. Because natural selection can act on groups of groups, and at many more levels, this phenomenon is now often called "multilevel selection."

WHEN TWO TRIBES GO TO WAR

Darwin's insight into group selection is seductive and simple. Take two rival tribes. One has members who are purely selfish and only out for themselves. The other has members who make a personal sacrifice for the good of their peers. Darwin thought that the latter tribe of "courageous, sympathetic and faithful members who were always ready to . . . aid and defend each other . . . would spread and be victorious over other tribes." As a result of group selection, they would thrive.

The first significant milestone in fleshing out Darwin's insight came with a book by the distinguished British zoologist and ecologist Vero Wynne-Edwards, known as Wynne to his friends. Wynne-Edwards argued that doing good unto others in the same group was "good for the species." In 1962, the Yorkshireman ignited an intense debate about group selection with his 650-page tome *Animal Dispersion in Relation to Social Behavior,* in which he suggested that animals are not, as Darwin supposed, always striving to increase their numbers but are instead programmed to regulate them for the greater good.

Those populations that showed restraint in reproduction and exploitation of resources survived longer than the more extravagant groups. Thus populations that produced fewer offspring could boost their chance of survival. Thanks to this kind of group selection, self-regulation of population size developed during the course of evolution. Like any good thing, group selection can be taken too far. Wynne-Edwards has been criticized for using this idea willy-nilly to interpret a vast array of social behaviors without proper evaluation.

It's true that some of his arguments were rather hand-waving. For example, let's say you observe that the males of a species fight and yet they don't kill each other. Why is this so? The kind of answer Wynn-Edwards would come up with is that if the males did kill each other, that would be to the detriment of the species. At a superficial level, this is not wrong. But it is sloppy and this problem demands a little more thought. Let me explain.

There will of course be variation among those males when it comes to fighting ability. It follows then that a male that wins a lot of fights—and kills—could have a selective advantage. If there is just one such rebel, then he, by definition, is more likely than his peers to survive and have offspring. Each generation that he sires will tend to inherit his selfish traits. After several generations of this natural selection, the gentler males in a group will be overrun by the killers.

This tension was not fully appreciated by Wynne-Edwards. In turn, his pioneering efforts were not appreciated by his peers. In that swinging decade, group selection became a pariah concept, taught primarily as an example of how not to think. One could speculate on whether the prevailing consensus of that day—that the good of the individual came above all else, even society—was influenced by the individualistic zeitgeist of the 1960s. And though Wynne-Edwards's book was the most controversial of its kind in its day, even its sternest critics would likely agree that it did at least help to stimulate this agenda of research.

During the intellectual brawl that Wynne-Edwards started, opponents of group selection punched several holes in the idea. The evolutionary biologist George C. Williams remarked in his 1966 book *Adaptation and Natural Selection* that "group-related adaptations do not, in fact, exist" and argued it was a theoretical implausibility because, in general, individual selection is a stronger force. The thinking at that time indicated that between-group selection was at best a weak influence on evolution when compared to within-group selection. Opponents also argued that the time elapsed between the overturning of generations in a group is longer than for individuals, so individual selection will occur much faster and so swamp any effects observed at the level of the group.

Most evolutionary theorists insisted that all adaptations had to be explained in terms of individual self-interest. They argued that the genes that do the best job of surviving to live on in the next generation, whether by crafty cooperation or by single-minded selfishness, are the ones that will eventually flourish. By this view, genes cannot become more frequent unless they benefit the individuals who carry them. That was that. A century after Darwin had sketched out his embryonic ideas on the matter, the phrase "group selection" was taboo.

Because of his dogged support for the idea, Wynne trod an increasingly lonely and isolated intellectual path. Nonetheless, he continued, as the Regius Chair of Natural History at Aberdeen University, to take every opportunity to convince the skeptics that group selection worked. He possessed a determined streak—he was also a tenacious cross-country skier and hill walker. In 1986 he attempted to answer his critics in *Evolution Through Group Selection*. He would, at the age of eighty-seven, write his last paper on group selection for the *Journal of Theoretical Biology*. He died four years later in 1997 at a retirement home near Banchory, where he had set up a team to carry out research on red grouse. Wynne spent his final months gazing out over the Dee Valley with his binoculars always at the ready.

Only a few scientists would carry the torch for the idea of group selection. Perhaps the best known is David Sloan Wilson, a professor at Binghamton University in New York. True to the belief that evolutionary biology took a wrong turn in the 1960s, he would work on the idea of group selection for more than three decades. This lonely knight was joined in his apparently quixotic quest by Elliott Sober, a philosopher of science at the University of Wisconsin–Madison, and later by the great naturalist Edward Wilson of Harvard (I will return to Wilson in chapter 8). To persuade their peers that between-group selection can be a match for within-group selection, they needed more than rhetoric. They required field studies and experiments.

These have begun to emerge. In the early 1980s, for example, David Craig of the University of Illinois, Chicago, conducted artificial evolution experiments with communities of flour beetles—pests that dine on wheat and other grains—in glass vials and found that group selection was effective. There's evidence of selection at the level of species too. David Jablonski, at the University of Chicago, has shown that over the past few hundred million years marine snails with limited geographic ranges have been more likely to go extinct than their more widespread kin. Proponents of group selection now required new models to bolster their experimental evidence.

MATHEMATICS OF GROUP SELECTION

My motivation to delve into group selection came, as ever, not from a conversation in a lab or outside a lecture theater, but on one of my walks with Karl Sigmund through the Rauriser Urwald in Austria. We both knew that there had been earlier attempts to model group selection, but we found none of them particularly convincing. They were too complicated for my taste. They did not give precise insights into the fundamental natural laws that would guide this potential mechanism for the evolution of cooperation.

In the heart of the Rauriser Urwald, Karl and I came across a little wooden board bearing a poem by Goethe (1749–1832). Goethe, one of my greatest heroes, was a polymath, a central figure in German literature who also wrote several influential scientific works, notably on color and development, and indirectly influenced Darwin. Goethe's poem begins as follows: "*Müsset beim Naturbetrachten immer eins wie alles achten.*" This translates as, "When looking at nature, you must always consider the detail and the whole." I could not think of a better way to express the idea of multilevel selection and, though I realized the field had a long, troubled, and vexatious history, I began to think about whether it might work in the real world.

The final impetus to roll up my sleeves and get to work on group selection came years later when my team was joined by Arne Traulsen, a thoughtful and witty physics student from northern Germany. We were looking for a project to do and I already had good anecdotal evidence that modeling group selection was his true calling. Arne seemed only too aware of the particular group that he himself belonged to. Hailing from Schleswig-Holstein, bordering Denmark, the instant that Arne crossed the Elbe River in Hamburg he felt he had left his home and entered that foreign territory that he disdainfully referred to as "Southern Germany."

When my conversations with Arne turned to group selection, I decided it was now or never. We began to develop a model, which we simplified every day until we had isolated the essence of multilevel selection. Our first paper appeared the next year, with the help of the physicist Anirvan Sengupta, and was further developed and refined the year after that. In these papers, we went back to basics. We threw out everything that was not essential for group selection to work, leaving behind only the mathematical kernel of truth.

We came up with the following simplified scenario, couched as ever in the language of cooperation and defection: When individuals interact with others in the same group, they get a payoff. The individuals can reproduce in proportion to payoff—so those who experience cooperation fare better than those who experience defection—and their offspring are added to the same group. Therefore cooperative groups grow faster.

In the second component of our model, we assumed that groups can break up, rather like cells divide or rival factions in a society part company to go their separate ways. By this I mean that if a group becomes too large, it can split into two groups. Occasionally, splinter groups also become extinct because there is only a limited amount of space (a creature runs out of habitat) or other resources (companies go broke for want of credit, for example). Thus we added the limitation that when one group divides into two, another group dies in order to constrain the total number of groups.

Run this model in a computer and you find that groups that contain fitter individuals reach the critical size faster and, therefore, split more often. What is satisfying is that this model leads to selection among groups, even though it is only the individuals that reproduce.

The higher-level selection of groups emerges from reproduction at the lower level of the individual.

Remarkably, the two levels of selection can even work against each other, when viewed in terms of cooperation and defection. Defectors can win within a group, but at the level of the group, groups of cooperators can triumph over groups of defectors. Thus, although cooperation might not always benefit cooperative individuals (whose efforts could be used and abused by cheats and free riders), we observed that groups in which cooperation emerges are more likely to stick around than those with exclusively selfish behavior.

Playing with the math of our model, a simple and compelling result emerges. Group selection allows the evolution of cooperation, provided that one thing holds good: the ratio of the benefits to cost exceeds the value of one plus the ratio of group size to number of groups. Thus group selection works well if there are many small groups and not so well if there are a few large lumbering groups.

Once we had sorted out the basics, we could elaborate the model to take into account other factors. For example, we could consider the effects of migration between groups. Migration makes it harder for cooperators to win because defectors might exploit and destroy one group and then move on to take advantage of another. As a result, migration undermines cooperation. We found that group selection still favored cooperation, provided that the benefit to cost ratio was boosted to compensate for migration. Our studies of the influence of migration have pointed to a simple conclusion. For efficient group selection, groups need mechanisms to prevent individuals from moving too freely between them and other groups: tribal loyalty or group stickiness, if you like.

To apply this new model to human behavior, one can start by thinking about genes. This may, for example, include genes that bolster the instincts of generosity, moral constraints, even religiosity. The genes that favor group cohesion, and discourage selection within groups, would also favor an innate sense of morality and group loyalty. By the same token, if conscience and empathy were impediments to the advancement of self-interest, then we would have evolved to be amoral sociopaths. But we have not. Groups that brim with men and women of mettle—brave, strong, innovative, smart, and noble—tend to prevail over those groups that lack integrity and grit.

A study by Sam Bowles of the Santa Fe Institute in New Mexico suggested that genetic differences between early human groups may indeed have been significant enough to account for the evolution of cooperation through lethal competition. Although there were few people in the late Pleistocene, 10,000 to 150,000 years ago, Bowles suggests that climatic swings at that time may have pushed once-isolated bands of hunter-gatherers into more frequent encounters. These encounters would have raised the possibility of conflict. When the threat of being wiped out by a rival group is sufficiently high, the costs incurred by individuals who make sacrifices for the good of the group can be offset by the increasing likelihood of survival of that group, including fellow cooperators.

To assess whether or not people with a genetic predisposition to cooperation could flourish through conflicts, Bowles took data on the lethality of ancient warfare and plugged them into a basic evolutionary model. The model pitted groups with genes for altruistic behavior against groups without. Without war, a gene imposing a self-sacrificial cost of a few percent in terms of lost reproduction would disappear from most of the population in 150 generations. However, he found that much higher levels of self-sacrifice could be sustained if warfare was brought into the equation. Overall, Bowles found that in many populations and for many plausible values of parameters "even very infrequent contests would have been sufficient to spread quite costly forms of altruism." In this way, he found support for the paradoxical theory that much of human virtue was forged and hardened in the crucible of war.

One can easily imagine how altruistic behavior can help insure a group against the costs of combat. For example, a broken leg could be fatal in a selfish group, since the injured member would be unable to forage and be at risk of starving. But food sharing in an altruistic group would mean an injured member could survive, ultimately making it less risky for that group to go to war. Sharing information and intelligence might have been another powerful way to protect the interests of your group and, by the same token, another powerful evolutionary force to encourage this kind of cooperative behavior.

Natural selection can act on human culture as well as on genes. Some scholars have scoffed that a scientific theory of cultural evolution that aims to ape Darwinian evolution is impossible because human beliefs and behaviors are so unpredictable and subject to historical contingencies as well as sudden breakthroughs, discoveries, and eureka moments. However, a neat way to illustrate how the forces of evolution can shape culture was put to work in research conducted in the Pacific by Deborah Rogers, Marcus Feldman, and Paul Ehrlich at Stanford University.

Their study examined the design of canoes from Fiji and ten Polynesian island archipelagos. The Stanford group catalogued a long list of canoe features, both functional and ornamental, and compared the rates of change from initial colonization of the remote Pacific three thousand years ago until the first European explorers encountered Polynesia. Historically in the Pacific, long sea voyages were essential for fishing, transport, and island life, and good or bad canoe design was a matter of life and death. A group that used a poor design would face a higher risk of extinction.

In all, the Stanford team examined ninety-six functional features— such as how the hull was constructed or the way outriggers were attached—that could contribute to the seaworthiness of the canoes and thus have a bearing on fishing success or survival during migration or warfare. They also evaluated thirty-eight decorative, religious, or symbolic features for comparison, covering a period from around 1595 to the early 1900s. Statistical test results showed clearly that the functional canoe design elements changed more slowly over time, indicating that natural selection weeded out inferior new designs.

GROUPS AT WORK

Until now, I have discussed groups of people. But, of course, an individual person could be considered a group consisting of a vast number of cooperative cells. Now the survival of the fittest individual boils down to the survival of the fittest cooperative of cells. By the same token, group selection works at the level of cells, and even at the molecular level. Individual cells that combine themselves into colonies or groups are analogous to replicating molecules that combine themselves into replicating cellular compartments.

Fascinating experiments have shown how compartments can compete. Harvard colleague Jack Szostak, a recent Nobel laureate, has suggested that at the origins of life survival of the fittest might have played out as a simple duel between fatty bubbles—early cells—stuffed with genetic material. Genetic material that replicated quickly may have been all that one particular primordial bubble needed to edge out its competitors and begin evolving into more sophisticated cells, according to experiments he conducted with Irene Chen and Richard Roberts of the California Institute of Technology.

These studies showed that genetic material could drive the growth of cells—little membrane sacs—just by virtue of being there. The RNA exerts an osmotic pressure on the inside of these sacs, which puts a tension on the membrane, which tries to expand. Growth is possible by stealing membrane from neighboring sacs that have less internal pressure as a consequence of having a smaller cargo of genetic material. In experiments, they saw that sacs swollen with genetic material grew, while those with no genetic material shrank. Thus the cells that had better-replicating RNA—and so ended up with more RNA inside— would grow faster. In this way their experiments showed that there is a direct coupling between reproduction—how well the RNA replicates— and how quickly the cell can grow. The successful cells sprouted hairs, and eventually a filamentous structure emerges that, with a shake, can spontaneously fragment into daughter cells. "That was a nice step," beams Szostak, "because it first showed us there could be competition between cells based on physical phenomena."

At the cellular level, there's plenty of evidence of group selection. Some of the most compelling data have come from studies of microbes. One strain of *Pseudomonas fluorescens* copes with a stagnant liquid medium by making a polymer that enables the microbe to form a mat on the liquid's surface. But the polymer is expensive to produce, and nonproducing bacterial cheaters can emerge that take advantage of the hard work of their peers. As they outcompete their brethren, the mat sinks. End result: at a group level, the polymer producing variety of bug is selected.

The effects of migration were revealed by experiments on groups of *E. coli* bacteria and bacteria-infecting viruses—phages—grown in ninety-six separate wells on plates. There were two kinds of virus: prudent phages, which are more productive when alone, and rapacious phages, which displace prudent variants when they share a colony of bacteria. Any rapacious phage mutant should oust its ancestor, but, of course, it will lower the overall productivity of its well and, as a result, run a higher risk of extinction. The success of the rapacious phage therefore depends on gaining sufficient access to fresh hosts to compensate for its lower productivity. When migration is unlimited, the less productive rapacious strain displaces the prudent one. But in biologically plausible migration schemes, prudent virus strains were able to outcompete more rapacious strains, despite their selective disadvantage within each group.

THE RISE OF MULTILEVEL SELECTION

Although the subject of group selection has been highly controversial, I believe there is now a wide range of evidence, both experimental and theoretical, to show that it is a distinctive and fundamental process that permeates all of evolution, from the emergence of the first cells to the behavior of social creatures such as humans. Group selection makes no assumptions about whether individuals are cooperative or selfish, let alone whether genes themselves are truly selfish. Instead, it simply says that intense between-group competition will favor mechanisms that blur the distinction between group and individual welfare if they improve performance or fitness at the group level.

For group selection to occur we need competition between groups and some coherence of groups. Different groups have different fitnesses, depending on the proportion of altruists. If 80 percent of a group is altruistic, it does better than a group that is enriched with only 20 percent altruists. So while selection within groups favors selfishness, those groups with many altruists do better. But, of course, the extent of group selection depends on important details, such as migration and group coherence. With this caveat, natural selection can indeed operate on several levels, from gene to groups of kin to species and perhaps beyond.

This leads to the Russian doll logic of groups within groups within groups. For that reason, as mentioned earlier, many people (myself included) also call group selection theory, multilevel selection theory. Focusing only on individual selection misses the bigger picture and overlooks crucial evolutionary processes that operate at higher levels. These are processes that take place among species, perhaps even whole ecosystems—that help to mold and to shape the world around us. The work I carried out with Arne provides the theoretical basis to explain how you can get an arbitrary number of levels of selection this way.

What is powerful about multilevel selection is that it works in a cultural sense as well as on DNA and genes. When we analyze the competition between neighboring tribes or between states, we need to consider both genetic and cultural forces if we are to understand why loyalty to a tribe, a church, or a neighborhood could override loyalty to the family, or naked self-interest. In other words, we need to model the "coevolution" of culture and genes. Even though group selection has a long and turbulent history, I believe that I have helped to establish that, in the right circumstances, multilevel selection is yet another mechanism that is responsible for the evolution of cooperation.

Like the mechanism of spatial selection that we encountered in the last chapter, the mechanism of multilevel selection can come to the aid of unconditional cooperators who do not use any complicated strategies. Understanding human evolution, and what paved the way for extraordinary behavior, such as the feat of bravery shown by Wesley Autrey, will require us to explore how group selection cooperates with direct and indirect reciprocity. What is remarkable is that there is a synergy between these mechanisms, so that the whole is greater than the parts. Multilevel selection can spark off direct and indirect reciprocity, blazing impressive new trails of cooperation.

CHAPTER 5

Kin Selection—Nepotism

I would lay down my life for two brothers or eight cousins.
—J. B. S. Haldane

Blood is thicker than water. The bonds of family and of common ancestry are often thought to be stronger than those of friendship and acquaintance. The more viscous the blood tie that links us to another person, the more we might strive to cooperate with them. This form of nepotism has evolved because we can increase the number of our genes passed to the next generation in this way and thus boost the size of our future genetic footprint. Known as kin selection, this is the fifth mechanism of cooperation.

The basic idea behind kin selection appeals to common sense. Cooperation can emerge more easily among closely related individuals. So far, so simple. A gene that induces you to cooperate with your brother or sister can spread by natural selection, because your relative—who is the recipient of your altruistic act—very likely carries the same gene. The literature on this mechanism is extensive. There have been some influential books, such as *The Selfish Gene* by Richard Dawkins, that have brought this idea to a wide and appreciative audience. Kin selection also has critics. When it comes to some recent developments in this field, I count myself among them.

Before I explain my position, let's start with the history. The development of this theory over the years weaves a fascinating tale. The fundamental idea behind kin selection was first glimpsed in a pub more than half a century ago by Briton John Burdon Sanderson Haldane. One of the biggest and most remarkable characters in twentieth-century science, Haldane is best remembered (along with Briton Sir Ronald Fisher and American Sewall Wright) as an innovative pioneer in population genetics, though he made many other important contributions.

Haldane was also a superb popularizer, writing hundreds of essays and books, from *Heredity and Politics* to *Daedalus; or, Science and the Future.* Evidence abounds in his writings of his laconic wit: "Intense selection favours a variable response to the environment. . . . Were this not so, the world would be much duller than is actually the case." In his collection of essays, *Possible Worlds,* Haldane came up with the immortal quotation "My own suspicion is that the universe is not only queerer than we suppose, but queerer than we can suppose."

Unsurprisingly, he was hugely influential. In 1924, when Haldane published his remarkable essay *Daedalus,* it was the first book to talk about the scientific feasibility of test-tube babies, which he called "ectogenesis." His work inspired Olaf Stapledon's *Last and First Men,* which charted the fortunes of humanity from the present onward across 2 billion years and eighteen distinct human species (which in turn motivated John Maynard Smith to take an interest in genetics and evolution). It also influenced Aldous Huxley's novel *Brave New World.* And in another novel by Huxley, *Antic Hay,* Haldane pops up in the guise of the character Shearwater, "the biologist too absorbed in his experiments to notice his friends bedding his wife." Haldane was also thought to have inspired the villain Professor Weston in C. S. Lewis's interplanetary trilogy *Out of the Silent Planet, Perelandra,* and *That Hideous Strength.* A Christian writer, Lewis fretted that slavish adherence to scientific materialism would erase idealistic, ethical, and religious values.

And so to kin selection. Legend has it that, one day in a pub in Bloomsbury in 1955, a lively conversation with Haldane over a few pints turned to the serious subject of the lengths that one would go to save another's life. Would, for example, Haldane risk his life to save a drowning man? Haldane, after a few moments' consideration, including some scribbling on the back of an envelope, replied, "No, but I would do it for two brothers or eight cousins."

In this way, Haldane built on and extended the sturdy idea that parents look after their offspring and thus their own genes. He had given the world an intoxicating insight into cooperation that would seduce and beguile generations of biologists: if genes were the key entities vying to get to the next generation, then it made sense for individuals to incur a cost if at the same time they conferred a benefit on relatives who carried the same genes.

This theory of kin selection acknowledges that a gene can propagate itself through two routes. The first is familiar—a gene can thrive by increasing the likelihood that the body in which it resides will survive and produce offspring carrying more copies. The second route is by increasing the reproduction of close relatives (kin) who also possess copies of the same gene. So brothers who share half of their genes are more likely to put themselves out for each other than cousins, who share one-eighth of their genes.

Haldane explained the idea as follows: "Let us suppose that you carry a rare gene that affects your behavior so that you jump into a flooded river and save a child, but you have one chance in ten of being drowned, while I do not possess the gene, and stand on the bank and watch the child drown. If the child's your own child or your brother or sister, there is an even chance that this child will also have this gene, so five genes will be saved in children for one lost in an adult. If you save a grandchild or a nephew, the advantage is only two and a half to one. If you only save a first cousin, the effect is very slight. If you try to save your first-cousin-once-removed, the population is more likely to lose this valuable gene than to gain it. It is clear that genes making for conduct of this kind would only have a chance of spreading in rather small populations when most of the children were fairly near relatives of the man who risked his life."

But this idea is not as intuitive as it might at first seem. Imagine there is a stranger before us, thrashing around in the foaming white waters of a turbulent river. Even though they are most certainly not brothers (or sisters) or uncles or cousins, I can call to mind plenty of people—think of the heroism of Wesley Autrey in the last chapter—who would still hurl themselves into the water without giving a second thought about relatedness. Haldane himself quipped: "On the two occasions when I have pulled possibly drowning people out of the water (at an infinitesimal risk to myself.) I had no time to make such calculations." There are other explanations for this act of altruism. Given what I have explained in earlier chapters, the mechanisms of indirect reciprocity and multilevel selection provide equally compelling explanations for this selfless behavior.

THE MATH OF KIN SELECTION

His short fuse and his unwillingness to suffer fools, made friendship hard for him, although for those of us who achieved friendship it was deeply rewarding. I loved him, but I sometimes wondered when he would discover that, by his standards, I was stupid.

—John Maynard Smith, Haldane's student

Haldane was a hero of Maynard Smith, who described how his inspiration had stuck to two guiding principles. One shining beacon was that "a physiological or biochemical explanation is more fundamental than a morphological one." His second was that "an ounce of algebra is worth a ton of verbal argument." When it came to kin selection, however, Haldane did not provide that ounce. The mathematical flesh and bones would be put on Haldane's insights in a doctoral thesis by Bill Hamilton. His workings were published in two long papers in the *Journal of Theoretical Biology* in 1964. Maynard Smith was one of the referees.

Hamilton started out his academic life as a shy and isolated figure at University College London in Bloomsbury. At that time, he seemed like a typical student—thin, shock-haired, soft-voiced, and unworldly. He preferred to work in his rented room in Chiswick, visiting libraries around London with a large canvas holdall stuffed with papers, books, a crumpled jersey, raincoat, and a fresh loaf. Sometimes he came to hate his solitary room so much that he abandoned it to work instead in Waterloo Station with, as he wrote, "the alcoholics there sheltering or craving company like me, the lovers parting, the fractious children herded by tired mothers."

The young Hamilton hungered to use mathematics to solve a major problem in biology, even though he had struggled with his "bible," Sir Ronald Fisher's classic text *The Genetical Theory of Natural Selection*. At the start of the 1960s he posted a card to his sister, now Mary Bliss, in which he confided: "I begin to think that my ambition to be a theoretical biologist can be more than a dream in spite of my poor mathematical ability." But then he became curious about altruism, from social insects to the warning cries of social animals. From the time of the early Greeks, for example, it had often been claimed that dolphins will save humans from drowning, or even defend them from shark attacks (Hamilton had come across such remarkable acts when asked to review a book on dolphins by his godfather).

Instead, Hamilton developed a mathematical formalism for the idea proposed by Haldane. He introduced the central concept of "inclusive fitness." The insight was an attempt to extend the concept of fitness, to recast it in a broader form. By fitness, biologists mean a measure of an individual's ability to survive and reproduce, the chance that one individual will leave more offspring in the next generation than others. The success of individuals with higher fitness is, of course, what leads to natural selection. When it came to the extraordinary degree of cooperation that is seen in social insects, such as ants and bees for example, Hamilton's theory of inclusive fitness suggested that this behavior can evolve because it still fulfills the selfish drive to pass on genes, though this mechanism now works through relatives instead of through the individual.

An animal may pass on its genes by helping its kin to reproduce, rather than by reproducing itself, because they share genes in common. Take the Belding ground squirrel, a small brown creature with a short tail, short fur, and rounded ears. When an individual makes an alarm call to warn of a looming predator, he puts himself in increased danger by giving away his location but helps protect his relatives and thus his genes. By being willing to risk sacrificing himself, the squirrel might allow for greater inclusive fitness. There's a fitness cost to the squirrel making the alarm call and a benefit to the receiver in the form of the chance to breed another day.

In a nutshell, Hamilton's formulation of kin selection says if the benefit, devalued by the relatedness between the two squirrels, was greater than the cost, genes responsible for such altruistic behavior could evolve. Hamilton's rule, which is written as $r > c/b$, is meant to predict just how much cooperation should be expected as a function of the degree of genetic relatedness. If the cost (c) of acting altruistically divided by the benefit (b) to the recipient of cooperation is less than the coefficient of relatedness (r) of the two individuals (the probability that both individuals possess the gene in question), then genes for cooperation could evolve.

Maynard Smith published ideas similar to Hamilton's at around the same time and coined the term "kin selection" (Hamilton always preferred his term of "inclusive fitness"). That same decade a "more graceful" version of Hamilton's mathematics—as Hamilton himself put it—would be developed by George Price, an American scientist who had been inspired by the 1964 papers. Price is a remarkable figure because, with no training in population genetics or statistics, he devised what came to be called the Price equation, a general statistical description of evolutionary change. As Hamilton himself remarked, the numbers popped out of Price's equation "like a rabbit from a conjurer's hat." Price believed the equation worked for group selection too and, though Hamilton found group selection woolly, he himself thought it could be made plausible with the help of Price's magical formula.

The Price equation did not, however, prove as useful as they had hoped. It turned out to be the mathematical equivalent of a tautology. The economist Matthijs van Veelen of the University of Amsterdam has a nice way to illustrate this point. He likes to use a quote by the famous soccer player Johan Cruyff, who won the European Cup three times, with Ajax, along with forty-eight caps—appearances—for Holland. Cruyff is famous for one-liners that hover somewhere between the brilliant and the banal. Some call them "Cruijffiaans." Matthijs

says the Price equation is akin to Cruyff's quip about the secret of soccer success, "*Je moet altijd zorgen dat je één doelpunt meer scoort als de
tegenstander.*" (You always have to make sure that you score one goal
more than your opponent.) Like the Price equation, it is certainly true.
But it does not get you very far if you are trying to understand the how
and the why of successful soccer playing. If the Price equation is used
instead of an actual model, then the arguments hang in the air like a
tantalizing mirage. The meaning will always lie just out of the reach of
the inquisitive biologist.

This mirage can be seductive and misleading. The Price equation can
fool people into believing that they have built a mathematical model
of whatever system they are studying. But this is often not the case.
Although answers do indeed seem to pop out from the equation, like
rabbits from a magician's hat, nothing is achieved in reality.

Price did however make a masterful contribution to the field. After
Maynard Smith referenced a paper that Price had sent to the journal
Nature in August 1968 (inspired by correspondence earlier that year
between Price and Hamilton), they joined forces to introduce game
theory analysis to the study of animal behavior, specifically to explain
why battles between members of the same species, such as Arabian
oryx, are ritualized encounters rather than serious fights. Maynard
Smith credits Price with first having the brilliant idea of extending
game theory from the traditional form pioneered by von Neumann—
rational decision making by brains when there are conflicts of interest—to decisions made by natural selection instead. This remarkable
1973 paper is a milestone that marks the beginning of evolutionary
game theory.

Price's biographer, Oren Harman, called him "genius, atheist-
chemist and drifter turned religious evolutionary-mathematician and
derelict." Bill Hamilton put it another way, which was no less intriguing. He recalled his old friend in a letter in which he explained how
Price's life, like a novel, was "exciting and unexpected right up to the
last page." The beginning of the last chapter of this novel came in 1970,
when Price had a religious revelation. Steve Jones, then a student at
University College London, recounts how he was approached by "a

stooped middle-aged American with a straggly beard and wild hair who told me, with great intensity, that he had a hotline to Jesus. Over the next few months his behavior became odder and he took to shouting in the corridors about his connection to the Savior."

Price moved from biblical scholarship to social work, often inviting homeless people to live in his house with him. He would become a tragic and disturbing tribute to the dangers of altruism in a society of defectors. Price gave up everything to aid alcoholics. Sadly, as he helped them, they stole his belongings. He eventually became homeless himself and ended up in an abandoned building in Tolmers Square, near Euston Road. By one account, Price was tormented by the feeling that he had failed to have a significant impact on easing human suffering. By another, he was troubled by the question of whether his altruism, indeed human goodness, was really genuine and pure. Is true, selfless altruism a fiction?

His torment soon came to an end. Price committed suicide sometime between January 5 and the morning of January 6, 1975. His body was identified by Hamilton, who had come to Price's squat to collect his meager effects. Hamilton vividly recounted his morbid duty: "As I tidied what was worth taking into the suitcase, his dried blood crackled on the linoleum under my shoes."

On January 22, 1975, as the rain fell, Hamilton attended Price's memorial in Euston. With him in the cemetery chapel a handful of tramps had gathered. Also present was Maynard Smith. At the end of the joint *Nature* paper that he had written with Price, the one that launched the field of evolutionary game theory, there's a poignant afterthought. A paragraph points out how a certain John Price had extended these ideas to "human neurotic behavior." John Price was a psychiatrist at the Maudsley psychiatric hospital in South London who had tried to help George when he was down and out. Today, George Price lies in an unmarked grave in St. Pancras Cemetery.

Hamilton himself would come to an untimely end. Richard Dawkins once described Hamilton as charmingly accident-prone. This was a reference to a succession of near misses: a childhood experiment with explosives that cost Hamilton several finger joints, his close encounters with Oxford motorists while pedaling madly on his bike, his decision to hike through Rwanda at the height of the civil war, and other hair-raising, sweaty-palmed adventures. Hamilton was a Felix with nine lives, said Dawkins. But his lives eventually ran out.

During the 1990s Hamilton had become increasingly convinced by the controversial (now dismissed) argument that the origin of the AIDS epidemic lay in the use of oral polio vaccines in Africa during the 1950s. Letters by Hamilton to major peer-reviewed journals were rejected, and to gather more evidence for this radical idea, Hamilton went with two companions to the depths of the war-torn Congo jungle. After a few weeks he was rushed back to London, apparently with severe malaria. This time, he didn't bounce back. Hamilton spent six weeks in Middlesex Hospital in London before dying on March 7, 2000, from a cerebral hemorrhage, aged sixty-three.

A great lover of insects, from fighting stag beetles to pharaoh ants, Hamilton wrote these poignant, evocative words a decade before his sad demise:

> I will leave a sum in my last will for my body to be carried to Brazil and to these forests. It will be laid out in a manner secure against the possums and the vultures just as we make our chickens secure; and this great Coprophanaeus beetle will bury me. They will enter, will bury, will live on my flesh; and in the shape of their children and mine, I will escape death. No worm for me nor sordid fly, I will buzz in the dusk like a huge bumble bee. I will be many, buzz even as a swarm of motorbikes, be borne, body by flying body out into the Brazilian wilderness beneath the stars, lofted under those beautiful and un-fused elytra which we will all hold over our backs. So finally I too will shine like a violet ground beetle under a stone.

His remains would never be consumed by Coprophanaeus beetles. A secular memorial service was held at the Chapel of New College, Oxford, and he was interred in Wytham Woods. He was laid to rest nearer his kin.

THE DECLINE OF INCLUSIVE FITNESS

One touch of nature makes the whole world kin
—Shakespeare,
Troilus and Cressida

I worked alongside Bill Hamilton in the zoology laboratory in Oxford in the 1990s. I admired him greatly and recall him as a kind and quiet person. I was always delighted when he sat down next to me during our afternoon tea breaks. That gave me a wonderful opportunity to try out new ideas on him and draw on his great depth of knowledge. By then, he was a pillar of the scientific establishment, a tall, striking figure with, as he put it, "two unrequitedly bushy eyebrows" and nostrils with "horse-hair bursts of an old Edwardian sofa."

When I studied in Vienna and Oxford I knew of kin selection from afar, like so many other theoreticians. I considered it to be an important theory that had a great deal of empirical support, in the form of studies of remarkably social creatures such as ants and bees. The theory had inspired many empirical biologists to measure relatedness in field studies and weigh up the costs and benefits of social interactions. At the same time, I found that the mathematical methods of kin selection were often murky. They were not as brilliant and crystal clear as the workings of Bob May in mathematical ecology and epidemiology, or the approaches used by Karl Sigmund in evolutionary game theory, or indeed as deployed in many of the great studies in traditional population genetics.

Equations seemed to arise out of nowhere in kin selection. There were many attempts at calculations that had no precise formulation of the underlying mathematical model—the equations used to capture and crystallize basic ideas. This is a recipe for disaster. Moreover, the concept of "relatedness" seemed to morph and change over time. The theoretical advances in kin selection began to drift away from other fields in biology, such as ecology, epidemiology, evolutionary game theory, and population genetics. As a result of these tectonic shifts, kin selection departed from the mainstream to become an island subculture with its own mysterious mathematical dialect.

I doubt that Hamilton would have approved of the way some of his followers have extended and diluted his idea of inclusive fitness. The most enthusiastic supporters of inclusive fitness theory began to proclaim it as a universal principle of evolution. For them, every aspect of the evolution of social behavior, from spite to cooperation, had to be expressed in terms of inclusive fitness theory.

I was struck by their attempts to rebrand rival ideas. Early kin selectionists had actually opposed group selection. But many modern kin selectionists have changed their minds. They consider group selection to be the same as kin selection. Neither position makes sense. Nor does the idea that "relatedness" is behind the evolution of all cooperation. Hamilton's rule became a dogma. Whenever the rule did not hold, the most fervent supporters felt they had the freedom to redefine cost, benefit, and relatedness for any new model in order to keep Hamilton's rule true. Eminent population geneticists—most notably Sam Karlin, Marcus Feldman, and Luca Cavalli-Sforza at Stanford University—had long ago pointed out the limitations of this approach.

Eventually I was moved to look more deeply at kin selection when I bumped into the great Harvard naturalist Edward O. Wilson, who was the very first person to support Hamilton's work. Lurking under his Southern charm and gracious manner is an appetite for intellectual brawls. Decades ago, Wilson had taken up cudgels to promote kin selection as a powerful explanation of "eusociality"—this is the term used to describe social insects and other animals that engage in the cooperative care of their young.

Indeed, kin selection was presented as a major principle in Wilson's 1971 book, *Sociobiology*, which put what he thought at the time was a beautiful and stunning concept on the map. The reason Wilson initially supported kin selection theory was the haplodiploid hypothesis, which I will explain in detail in chapter 8. Wilson's book proved highly influential. We can tell this because most early citations of Hamilton's work use the incorrect reference that Wilson himself mistakenly put into *Sociobiology*. By the same token, it is clear that at that time few people bothered to go back to Hamilton's original papers. Over the years, as evidence for this hypothesis steadily crumbled, Wilson began to harbor serious doubts.

When I first discussed kin selection with Wilson, we both made a fascinating discovery. Wilson told me that he always thought kin selection was a great mathematical theory but added that, since his greatest infatuation with the idea in the 1970s, he had become increasingly disillusioned because the real-world evidence had, if anything, become weaker. Many studies have come to the conclusion that, as a rule, members of social insect colonies cannot actually recognize their own degree of relatedness to their nest mates. Returning to Haldane's dilemma, if you are unable to distinguish a brother from a cousin then it is hard to see how kin selection can possibly operate at all. That was striking. I told Wilson that, in contrast, I had always had the impression that the empirical support for kin selection among social insects was strong. My problem was that the mathematics was obscure. In this way, our encounter led to mutual liberation.

We began to meet regularly. We were joined by Corina Tarnita, a remarkable mathematician whom we will meet again in chapter 13. Corina possessed the mettle and patience to work through all the calculations that had been performed in kin selection theory. Over a year, she hacked her way through an impenetrable jungle of math and biology. Every morning, she awoke optimistic and hopeful that she would find something sublime in these calculations, a mathematical structure of great simplicity and beauty. She never did. But she did make an amazing discovery, marking a paradigm shift away from the idea of inclusive fitness and back to the old-fashioned concept of natural selection.

The fundamental idea at the heart of the kin selection jungle is Hamilton's concept of inclusive fitness. So let's begin with Hamilton's own definition: "Inclusive fitness may be imagined as the personal fitness which an individual actually expresses in its production of adult offspring as it becomes after it has been first stripped and then augmented in a certain way. It is stripped of all components which can be considered as due to the individual's social environment, leaving the fitness which he would express if not exposed to any of the harms or benefits of that environment. This quantity is then augmented by certain fractions of the quantities of harm and benefit which the individual himself causes to the fitnesses of his neighbors. The fractions in question are simply the coefficients of relationship appropriate to the neighbors whom he affects; unit for clonal individuals, one-half for sibs, one-quarter for half-sibs, one-eighth for cousins, . . . and finally zero for all neighbors whose relationship can be considered negligibly small."

We have to contrast Hamilton's idea with the standard approach of natural selection. In every model of evolutionary dynamics (be that in ecology, evolutionary game theory, or population genetics) we usually calculate the fitness of individuals by taking into account all the necessary interactions that occur in the population. This common sense approach calculates the fitness of every individual in the population and then evaluates how this affects the survival of genes that induce certain behaviors, such as the urge to save drowning siblings. Then we can see whether a genetically encoded strategy, such as saving a brother or saving a stranger, is favored by natural selection or not. This is the normal approach based on "fitness" and "natural selection," one that Haldane himself would have adopted.

But inclusive fitness theory proposes another approach. Think of it as an alternative way of accounting, if you like. Rather than calculating the fitness of individuals, the proposition is that we only consider one point of view, that of the "actor," say the brave person who leaps into the turbulent, foaming waters. We calculate how his action affects his own fitness and the fitness of the two brothers, the eight cousins, or whoever happens to be drowning in Haldane's treacherous river. These are the recipients. But if the saved relatives compete with other relatives for reproduction, then their fitnesses have also been affected and need to be factored into the inclusive fitness calculation. We see that inclusive fitness is by no means a simple idea. Finally we add up these fitness components and multiply them with the relatedness between actor and recipient. The resulting sum is called the inclusive fitness of the actor. Note that it only includes what comes as a result of the actor's own actions, not as a result of the help received from others.

It must be clear by now that this kind of accounting is not straight-forward. Normally, it is not possible to deconstruct the fitness of an individual into additive components. In general, fitness is a more complicated function of what others are doing, and then it is not possible to calculate inclusive fitness. For this reason, it is not at all a universal or a robust idea, but rests on particular mathematical assumptions.

Let me describe some of these assumptions. For one, inclusive fitness can only be defined when the fitnesses of all individuals in the population are almost identical, with vanishingly small differences. It only works at what biologists call "near neutrality," where there is almost no selection at work. But, we discovered, even if we are at this limit of near neutrality, inclusive fitness theory does not always work and we still need more conditions.

Among the restrictions required for inclusive fitness theory to work, one is that all interactions must be additive and pairwise. But in an ant colony, for example, many tasks require the concerted action of several individuals, not just pairs. Think of the transport of heavy food items, or the defense against attack by another ant colony. Moreover, inclusive fitness theory only works for very simple, somewhat static population structures. For a formulation of an inclusive fitness calculation, all these assumptions have to be made. By the same token, if these assumptions do not hold (and, of course, they rarely do), then the concept of inclusive fitness has no meaning.

We made another discovery. Let's give inclusive fitness theory the benefit of the doubt. We will assume, for the sake of argument, that we are in the happy situation in which all these highly restrictive assumptions hold true. We are now in a utopia called "inclusive fitness land," where individuals working in pairs are part of a special population structure and where selection hardly operates at all. We found that in this special world where inclusive fitness theory works, the calculations yield up exactly the same prediction as standard natural selection theory. Hence inclusive fitness theory comes up with no novel predictions or insights. Casting a problem in terms of inclusive fitness is like having to undergo elaborate and time-consuming initiations to join an elite club, only to end up with nothing in the way of privileges.

This now begs the question: If we have a theory that works for all cases (good old-fashioned natural selection theory, which is also focused on genes) and we have a theory that works only for a small subset of conceivable cases (inclusive fitness theory), and if for this small subset of cases the two theories lead to identical results, then why not stay with the "easier" and more general theory?

But let's give kin selection one last chance. Perhaps the most compelling insight to arise from the theory was Hamilton's rule. This was his recipe for altruism that rested on the idea that you can die for close kin and still spread your genes, since close kin have many genes in common with you. The rule states that cooperation can emerge if "relatedness" is greater than the cost-to-benefit ratio, predicting just how much cooperation should be expected as a function of the degree of genetic relatedness. The problem is that even if we are in the utopia where the inclusive fitness assumptions apply, the simple form of Hamilton's rule, shown above, does not hold. Inclusive fitness theorists are aware of this problem and have tried to redefine the cost and benefit parameters in Hamilton's rule in order to make it work. But this form of repair comes at a heavy cost: you lose predictive power and mathematical insight.

There is also a problem with the experimental tests of inclusive fitness theory. If empiricists measure genetic relatedness in their laboratory experiments or in field studies (often a very useful measurement that tells them about population structure), then they are not testing inclusive fitness theory. In order to do this they would actually have to set up an inclusive fitness equation, which takes into account detailed and complex aspects of population structure and weighs up the costs and benefits. To my knowledge, these extensive measurements have not been performed.

It is often remarked that inclusive fitness theory is a gene-centered approach. In reality, however, it is centered on the individual. Let us consider, for example, an ant colony with queen and workers. This lifestyle is an example of eusociality: there are workers who do not reproduce but instead help the queen to reproduce. Inclusive fitness theory makes the worker the center of attention, asking why the worker behaves altruistically and raises the offspring of the queen. Why doesn't the worker ant leave the colony, mate, and raise her own offspring? To answer this question, kin selectionists believe they have to go beyond natural selection and analyze the inclusive fitness of an individual's action, such as a worker ant helping the queen raise her offspring. They believe that before the invention of inclusive fitness there was no satisfactory answer to this question. I beg to differ.

If, instead, you develop a model that is truly gene-centered, you realize there is no need to use the concept of inclusive fitness at all. All we have to ask is whether a gene linked with social behavior wins out over one linked with solitary behavior. Now the central question becomes the following: Is that social gene favored or opposed by selection? The calculation that is needed to answer this question is entirely based on standard natural selection theory.

In this way, Corina Tarnita, Edward Wilson, and I have developed a novel, alternative way to explain the evolution of eusociality without resorting to the gyrations of inclusive fitness. We use natural selection combined with a careful consideration of population structure to explain the evolution of the extraordinary cooperative societies of insects such as ants. We can reach satisfactory conclusions without any need for inclusive fitness at all. I will return to this work in chapter 8.

You may think, given my somewhat withering analysis of inclusive fitness, that I think kin selection is dead. This is not the case. Despite its limitations, Hamilton's rule has been a valuable heuristic. Inclusive fitness thinking has inspired much theoretical work over the years and led empirical biologists to measure relatedness in many field studies. But in my opinion it is time to move on. It is now clear that the mathematical understanding of evolutionary dynamics has advanced to the point where the theory of sociobiology can be extended beyond Hamilton's rule. We are now at the point where we can ask more detailed, more accurate questions and achieve a much deeper understanding of evolutionary processes.

Kin selection is still a mechanism for the evolution of cooperation, as long as it is properly defined. I can envisage it working whenever there is conditional behavior based on kin recognition. So, depending on whether I am looking at a brother or a stranger, I will behave accordingly. By conditional behavior, I mean that I may throw myself into a river for my brother but not for a stranger. What I find telling is that, more than half a century after Haldane's famous quip, details of those circumstances have still to be worked out. If kin selection is to move forward, we have to go back to Haldane's original sparkling insight.

Feats of
Cooperation

CHAPTER 6

Prelife

There is grandeur in this view of life, with its several powers, having been originally breathed into a few forms or into one; and that, whilst this planet has gone cycling on according to the fixed law of gravity, from so simple a beginning endless forms most beautiful and most wonderful have been, and are being, evolved.

—Charles Darwin, *On the Origin of Species*

Centuries ago, the Scottish botanist Robert Brown became fascinated by the zigzag motion of fragments within pollen grains. In his pioneering observations, made with a primitive microscope, Brown had spotted this random jittery motion as early as 1827. What puzzled him was that this incessant movement did not arise from currents in the fluid, or from evaporation, or from any other clear-cut cause. He was aquiver with the idea that he had glimpsed an animating force— "the secret of life."

Being a good scientist, he knew that he needed more evidence. At that time, Brown was well known and had been honored across Europe, gaining an honorary doctorate from Oxford in the same ceremony as the great physicist Michael Faraday and chemist John Dalton, pioneering the exploration of Australia, and even advising Darwin on what equipment he should take with him on his famous voyage of discovery aboard the *Beagle*. After observing the same kind of motion within mineral grains, which were obviously inanimate, Brown discarded the idea that he had seen the vital essence at work, with understandable disappointment.

Yet, in a certain sense, he had indeed glimpsed an animating force. The key step toward making sense of what Brown had actually witnessed came more than seventy-five years later, when Albert Einstein demonstrated how the tiny zigzagging particles were being buffeted about by the invisible molecules that made up the water around them. The existence of molecules was still rejected by some major figures in the scientific establishment of 1905. Einstein predicted that the random motions of molecules in a liquid putting pressure on larger, suspended particles would result in the particles' irregular motions, big enough to be directly observed under the gaze of a microscope. From this jittery motion, Einstein could even work out the dimensions of the molecules. Although Brownian motion did not turn out to be a vital force, Brown's observation paved the way for the understanding that we now use to explain early life.

Living, breathing creatures did not require a life force, or vital essence, to evolve but an extraordinary level of cooperation between molecules. Some of these molecules originated in the atmosphere of primitive Earth, split and fused under the action of ultraviolet and other radiation to form simple organic molecules, such as hydrocarbons. Lightning strikes provided a source of high energy that could extend the repertoire of molecules. Hydrocarbons provided the feedstock for more complex organics such as simple amino acids (which, when linked together, form proteins) and carbohydrates (simple sugars). Somehow, these molecules became organized into bodies, the ancestors of cells, with definite shape, unity, and properties that resembled those of living things.

Look under the hood, into the cells of the life that thrives around us, and you will find master molecular replicators, the messengers of our inheritance. Today, the most common is deoxyribonucleic acid, DNA. Apart from a few viruses, all life on Earth now relies on DNA to hold the information that it needs to reproduce. But the most likely players in the first games of life were molecules of the related genetic material RNA, which is more flexible than DNA because it can both carry information down the generations and also catalyze—speed up—chemical reactions, a very handy feat. And RNA still carries out all kinds of critical roles in organisms that are described by DNA, including human beings. In 1986, the Harvard Nobel laureate Walter Gilbert coined the term "RNA world" to suggest how RNAs could have dominated the story of our origins, before proteins entered the game of life.

A visit to an island paradise would inspire me to take this quest to understand our origins in a new direction. Good science is about asking the right questions, and it turned out that a rephrased version of the one about life's origins germinated the seed of an intriguing new idea in my mind. Conventional thinking says that organisms that are more successful go on to reproduce more and pass their genes on to more offspring. Thus reproduction comes first, and then selection. So far, so reasonable.

We also know that evolution crafted all life on the planet with natural selection and that we can capture the way it works with mathematics, distilling its essence into the form of equations. In this way I devised a general mathematical theory, not for evolution itself but for the *origin* of evolution. Working with colleagues, I showed how, in its infancy, Earth generated a complex ecosystem of cooperating molecules. Over millions of years, this formed the molecular equivalent of dry tinder for a coming firestorm of replication that we now know as life.

I concluded that the process of natural selection, which is linked with the vast diversity of life from bacteria to tigers, actually predates the emergence of reproduction itself, that is, the ability to make copies whether in the form of eggs, offspring, or molecules. By thinking through the implications of this theory about life's origins, I could also show that cooperation is more ancient than life itself.

PASSAGE TO PARADISE

The phone rang one day, when I was at the Institute for Advanced Study in Princeton. Within a minute or two I found myself explaining my research to a stranger who had introduced himself as Jeffrey Epstein. He turned out to be a Wall Street tycoon. The next day, his office wired my administrator a generous donation to fund my research. Later came an invitation to visit him in New York, and I found myself in a former school that had been converted into a magnificent mansion. I had been invited for dinner and I was flattered to find that I was the only guest. We talked for hours. He loved my work on cooperation and wanted to know every detail. He was particularly intrigued by the strategy of Win Stay, Lose Shift. He often challenged my perspective with new ideas. It was an engaging discussion and would be the first of many.

Epstein wanted me to organize a conference on the evolution of language. This event took some planning and would eventually take place a year later at the Institute for Advanced Study. Epstein himself turned up at the start of the gathering. His private jet was on standby at Princeton Airport and, even though he would make a quick getaway to Paris after a short while, the meeting seemed to whet his appetite for my research. Some time later, he invited me to visit him again.

A female member of Jeffrey's household rang to make the arrangements. There would be a ticket to fly me to San Juan, Puerto Rico. From there, I would be picked up by helicopter. She casually added that she would be the pilot. Now I felt like an extra in a James Bond movie. Not long after, I found myself overlooking warm, cobalt waters. I was sitting at a stone desk in a colonnaded courtyard on a speck of land in the Caribbean. Jeffrey's tropical island consisted of only 110 acres in all and was ringed with reefs. There was a luxury retreat with patina-finished roofs, and a mile-long beach dotted with palm trees that had been imported from Florida. To warn off pirates, a star-spangled banner fluttered high in the wind.

My guesthouse had blue shutters and the interior had been decorated by artists who had been flown in from France. Several of these little houses were arranged around a fountain, courtyard, and kidney-shaped pool. There were sofas and lounge chairs dotted about, but I always liked to work at the stone table. Every day I breakfasted with Jeffrey as the sun rose. We would have endless conversations about science, about my work, what it meant and where it was headed.

Jeffrey was the perfect host. I asked a casual question about what it was like to dive in the warm, clear waters around the island. The very next day a scuba diving instructor turned up. When the British cosmologist Stephen Hawking came to visit, and remarked that he had never been underwater, Jeffrey rented a submarine for him. On the last day of my visit, Jeffrey said he would build an institute for me.

In 2003, after negotiations between him and our then president Larry Summers, I was able to set up the Program for Evolutionary Dynamics, PED. Summers, who went on to become the chief economic adviser to President Obama, gave me some blunt advice on how to do this within limited resources: "Spend it. There will always be enough money." We occupied the top floor of a smart new office block in Brattle Square in Cambridge, a central spot surrounded by restaurants, boutiques, shops, and during the summer, buskers and street performers. There I could be joined by a hand-picked group of crack mathematicians, biologists, or indeed anyone who wanted to explore the remarkable power of cooperation. Some students called my program "Nowakia." For us, Nowakia was an academic paradise. One of Harvard's most impressive graduate students, Erez Lieberman, joked that PED could also stand for "party every day."

Depending on visitors, I had between fifteen and twenty-five people at any one time investigating all kinds of cool problems. There were plenty of undergraduate students too. We called them "Romans," because they were formally part of a program called Research Opportunities in Mathematical Evolution. Erez Lieberman styled himself as Decurio, the name given to the leader of a Roman squad of ten people, and used it as an excuse to crack toe-curling jokes. If too many students showed up to work with us, Erez would announce that "all roads lead to Rome." If a student's research project had run into difficulties, he would comfort them with the words: "Don't worry, Rome was not built in a day."

My empire sounds like something out of an old Marx Brothers movie. I aspire to be its Rufus T. Firefly, the legendary dictator with the statesmanship of Gladstone, the humility of Lincoln, and the wisdom of Pericles. In fact, I have a simpler objective: I want the inhabitants of PED to be fulfilled in their quest to understand nature, and happy too. I want them to have options, not obligations. They don't work for me. I work for them.

THE PROBLEM OF LIFE

On his island paradise, Jeffrey had plenty of time to think. Several years later as we both sat at the same stone desk, he returned to one of the biggest questions of all: "What is life?" Yet he put this Biggest of Big Questions in a more interesting way. He added: "Life is the solution. But what is the problem?" After all, the orbits of the planets around the sun are the answers to how these celestial bodies "solve" Isaac Newton's equations of gravity. The movement of electrons around the nuclei of light atoms "solves" equations of quantum mechanics. Which equations does life "solve"?

I loved the way he phrased it. Too bad I did not know the answer. I dodged his question. Life, I said, is that which evolves—it solves the equations of evolution. But that begs the question "What is evolution?" This is another Big Question because, of course, biology is evolution. Then I realized that the key issue is actually this one: How does evolution begin? We know that living things can evolve. Evolution transforms one living system into another. But how does evolution itself come into existence? What was there before evolution? This seemed to be at the heart of what we wanted to discover.

What spirit in the void moved a soup of dumb molecules of the kind that once entertained Robert Brown to become a smart, living, organized biochemical brew like Brown himself? Fired up by the possibilities, I immediately began to develop a scheme of chemical reactions where I could see a gradual transition from nonlife to life, from pure "nonliving" chemistry to biology. I imagined how two chemical units could be knitted together—polymerize—to form sequences. One could think of the two subunits as the 0 and 1 of a binary code. I could see in my mind's eye how this polymerization reaction could explore a universe of possible codes with these simpler binary subunits. In this way, while sitting at that desk in paradise, I came up with a new concept, that of "prelife."

Here's how I began to formulate the problem. I used simple units to represent the alphabet of the first chemical building blocks of life, whether RNA or whatever kind of molecule happens to be a fashionable candidate. These basic building blocks randomly and spontaneously assemble into strings of information, just as letters spell out words. I was interested in the rate at which this happened at the dawn of life. Or, to put it another way, my study focused on the kinetics, how quickly strings with different sequences will grow. Strings that encode different kinds of information will become longer at different rates, with some taking in chemical building blocks faster than others. Small differences in growth rates result in small differences in abundance.

The mathematics shows that because longer chains require more assembly reactions, they should be less common than short chains. But if some assembly reactions run faster than others, then chains built from these fast-assembling sequences of building blocks grow to be larger—just as cranking the handle of a pasta maker more quickly will generate longer pieces of pasta. This triumph of spaghetti over macaroni is a form of selection. By looking at the primeval pasta soup this way, I could see that it was possible to have selection prior to replication, and it emerges in a natural way.

Some strands may also mutate (think of pasta with different shapes, some tubular, others twisted, and so on) and the new mutated progeny may grow more successfully. Sometimes one strand accelerates the reaction rates of other sequences, marking a form of cooperation. When taken together, the possibilities added up to form a lifelike chemical system that is pregnant with the potential for evolutionary dynamics. In this way, molecules subject to the forces of selection and mutation that are themselves still quite incapable of replication—the final condition necessary for life—are drawn inexorably toward it.

So far, I have not considered replication, only the reactions required to make basic molecules. Now if some do have the ability to replicate, then we can ask whether there is selection for replication or not. In this way, we can ask about the precise conditions for life to emerge out of prelife. Simpler units—monomers—lead to prelife. Prelife builds life. But then there is competition between prelife and life for the basic feedstock of monomers. This competition could lead to the selection of molecules that are able to replicate.

This change of emphasis, from the origins of reproduction to those of selection, has a dramatic implication. The standard view is that some 4 billion years ago our planet witnessed a singularity that marked the ultimate birth of the living diversity that can be seen around us today, a big bang in biology that forged the first link in a chain of reproduction that stretches from the moment a band of cooperative molecules crossed the watershed between inanimate chemistry and biochemistry to evolve into all the things that wriggle, swim, crawl, and walk.

In the traditional picture of the origins of life, the instant the first replicator pops up is a rare and fleeting moment that bootstraps the rest of evolution. This is an enormously lucky event, a fleeting spark that lights the fuse of all life. As in Greek cosmology, from a "vast and dark" void, chaos, was born Gaia, the primordial Earth goddess. But the work on prelife suggests there may never have been such a magic moment—no singularity and no Big Biological Bang. Instead, I could see a gentle transition, a smeared boundary between chemistry and biochemistry, between "prelife" and life. Life had no definitive beginning but gradually came into vibrant focus from a blurred, dark origin.

Over the eons, prelife became enriched. The richer prelife became, the more likely it was that an outbreak of life would take hold. In other words, the development of rich chemistry—with enough time and space—was bound to discover the right molecules to replicate. In this way, the soup of lifeless chemicals on Earth was, in effect, testing possible replicating molecules, and making it much more probable that one might eventually reach the threshold of life.

Thus, the origin of life may not have rested on a single spark of genesis but on a chemistry that churned on Earth for hundreds of millions of years. Within the chemistry of prelife, there were opportunities for cooperation. Some sequences could have had catalytic activities, which means they increased the rate of certain prelife reactions. I could also envisage how two complementary prelife sequences could have catalyzed reactions which built each other. One molecule increased the rate at which the other was formed, and vice versa. The existence of cooperating pairs of molecules in prelife is very plausible. Indeed, replication of a single strand of RNA can be thought of this way: one strand of RNA builds a complementary strand, and so on. Thus cooperation is older than life itself.

So here you get a more nuanced view of what happened at the dawn of biology. A few strings of basic units eventually develop the ability to make copies of themselves, and reproduce, if there are enough chemical units around. The replicating strings could eat up the basic units faster than the nonreplicating strings, so there is competition between prelife and life itself. By our calculations, only when conditions arise that tip the rate of replication over a certain threshold would prelife be overwhelmed by replicating strings. Ultimately, life gnaws away the scaffold of prelife molecules that got it going in the first place. In this way, life is an infection of prelife, one that eventually destroys its molecular ancestor. Or to put it in an anthropomorphic way, life exploits prelife.

There are even wider implications to this idea, which are worth a tiny diversion. It is tempting to speculate on what will happen as computing environments become ever richer, and as the internet connects ever more powerful computers dotted around the planet. One can envisage the day when computer-based life forms—evolving software, or soft-life—manage to evolve spontaneously by a process very similar to the one that first paved the way for life on Earth 4 billion years ago.

WHAT NEXT?

Once you have a population of replicating strings, say of the genetic material RNA, what follows? Much work has been done on developing mathematical equations to describe how the concentrations of molecules, such as RNA polymers, can change over time in chemical reactions. The rate of self-replication can vary according to the sequence of genetic "letters" in the RNA polymer: some sequences of RNA may produce "offspring" faster than others. In other words, they are "fitter."

But, of course, replication is error prone and this must be taken into account. Because of the never-ending inaccuracies in reproduction, an offspring sequence need not be identical to its parent. In this way, random chemical events during reproduction can create a spectrum of RNA molecules with different sequences, albeit related. This ensemble of closely related RNAs is called a quasispecies. The consequences of having a rainbow of replicating molecules present at the outset of life was studied by Manfred Eigen, a German Nobel Prize–winning chemist, together with Peter Schuster of the University of Vienna.

To begin with they needed a way to envisage the possibilities for evolution. In 1932, the American geneticist Sewall Wright had come up with the idea of a "fitness landscape." This concept was applied by Eigen and Schuster in an extended form to describe RNA evolution and the origin of life. First they built what is called sequence space. They arranged all RNA strands of the same length in a lattice such that neighbors differ by only one chemical letter, or base. Now the distance between any two sequences equaled the number of mutations (differences in letters) between them—the more mutations that separated the sequences, the further they were apart.

Each RNA sequence has a certain rate of reproduction, which is the "fitness" of that sequence. Sequences with relatively high fitness reproduce faster and become more abundant as they outcompete slower, less fit sequences. If the fitness of each particular RNA is drawn on a vertical axis over sequence space on a horizontal axis, then we end up with the fitness landscape that characterizes RNA evolution.

Bearing all this in mind, we can imagine how faster-replicating sequences form mountain ranges, with the fastest of all represented by the tallest peak. On the landscape, the slowest-replicating sequences languish deep in the valleys below. And where neighboring RNAs have a similar fitness, there are pancake-flat plains. Now the quest for the best replicator can be expressed as the hunt for the most dizzying peak in a vast, rolling landscape of possibilities.

An evolving population typically clambers uphill in the fitness landscape, by a series of small genetic changes, until a peak is reached. One of the subtleties of evolution is that this particular peak may not be the tallest, and so a population may be marooned there until a rare mutation opens a new path to an even higher fitness peak. Otherwise, the route to a taller peak may pass through a valley, the quasispecies equivalent of that old adage that things will have to get worse before they get better.

The landscapes are multidimensional. If you try to draw such a lattice for binary sequences of lengths two and three, you will need as many dimensions as the sequence is long. To draw the sequence space of the human immunodeficiency virus, you would need 10,000 dimensions. For the entire genetic complement of a human being, the human genome, you need about 3 billion dimensions. Compared with this, the strands of genetic material at the dawn of life were relatively short, though still beyond the ability of a person to envisage the multidimensional space that they inhabited.

I hope that you can begin to appreciate how the possibilities of evolution dwell in very lofty multidimensional spaces. These are actually more "spacious" than the universe itself. They are so vast that somewhere in these high dimensional Himalayas, amazing Shangri-las of peculiar and exotic life-forms may exist in a fitness landscape as oases that evolution will never discover in the lifetime of our cosmos.

PLAYING THE GAME OF LIFE

When Eigen and Schuster introduced quasispecies theory in the late 1970s, they created new opportunities to understand life and evolution. Quasispecies can roam far and wide over the high dimensional fitness landscapes searching for peaks, which represent regions of high fitness values. But Eigen and Schuster argued that the target of natural selection is not the fittest sequence but the fittest quasispecies. This is an important distinction because the fittest sequence may only represent a very small fraction of the quasispecies. Indeed, it may not even be present all the time. Moreover, the highest peak may not correspond to the fittest quasispecies, as will become clear.

It is easy to see why the idea of the quasispecies, a grouping of slightly different but related RNA molecules, is so powerful. You might think that the success of any single RNA replicator depends only on its own ability to make itself and thus on its own rate of replication. But this is not the case: it also depends on the replication rate of nearby mutants. The reason is that, after the right mutation, these nearby mutants can also generate the original RNA replicator. the different close-by sequences can "cooperate" with each other via mutation.

In this way, natural selection picks the fittest cloud of RNAs (quasispecies) rather than the fittest RNA sequence itself. The consequences can be counterintuitive when it comes to the game of life. Imagine two RNA sequences, A and B. Let's assume that A has a higher replication rate than B, thus it has a higher fitness value. The conventional view is that A should win the game of life. Correct? Not so fast.

Let's suppose that A is surrounded by mutants with very low fitness, so it is atop a sharp peak, while B is surrounded by mutants with relatively high fitness, more like a tabletop mountain. The tabletop mountain is at a lower altitude than A's narrow peak. Play with the equations, as Peter Schuster did at the University of Vienna with Jörg Swetina, and you find that as the mutation rate increases A loses the competition with B. There is a critical mutation rate, below which A is the winner, but above which B and its neighbors are favored, and this can be worked out with the "quasispecies equation."

There are other important consequences of quasispecies theory. Biology students have it drummed into them from day one that mutations are random. They are not directed. They are blind. But, as we have seen, evolution is most certainly not. By this I mean that the mutants most suited to an environment tend to prosper. Now we can see that selection can act on the structure of the quasispecies and guide it along ridges to the nearest peaks. This happens because more successful mutants produce more offspring than less successful ones (which are further away from the peaks).

Before this insight, one might have been tempted to think of evolution as a completely random walk across a fitness landscape, in other words a series of steps in multidimensional sequence space where no region is more likely to be explored than any other region. But it turns out that, due to the power of selection and quasispecies, there are biases in evolution's walk through the genetic possibilities of life.

ERROR LIMITATIONS

The extraordinary possibilities of evolution are the result of errors in replication. Mutations are critical. If the replication of RNA were error free, and perfect, no mutants would arise and evolution would stop. Nothing would change. There would be no living diversity. So mutations are required for life to emerge. Equally, however, evolution would be impossible if the error rate of replication was too high. The reason is that only some mutations lead to an improvement in adaptation. Most lead to deterioration.

By playing with the numbers, one finds that if too many mistakes occur in any one replication event, a population of RNA molecules will be unable to maintain enough meaningful information to pass to the next generation as the genetic message becomes scrambled. So when the mutation rate exceeds a sharply defined threshold, inheritance breaks down. Thus if a quasispecies of RNAs is suited to work well in an environment, once it moves past this threshold of error, any adaptation to that environment is impossible. Eigen and Schuster found that there was a way to figure out where this error threshold lies and express it as a maximum-possible sequence length that can thrive for any given mutation rate.

Let's say that there is a given probability of a spelling mistake when an RNA sequence reproduces. Naturally, the longer the RNA sentence, the more errors it will contain, just as the longer the sentence you attempt to spell out by hand the more of a chance there is of making a silly slip. So we can think of the error threshold as a length of RNA past which the ability to pass on genetic information is too degraded. For an RNA of 100 "letters" (chemical units called bases) the mutation rate has to be fewer than 1 in every 100 letters for the message to be passed down. And for an RNA of 1,000 letters, it has to be fewer than 1 in 1,000. Thus the maximum possible mutation rate per base must be less than the inverse of the genome length. That way, there will be enough descendants with the correct message to pass the information down the generations.

Experiments by Leslie Orgel at the Salk Institute in San Diego on the spontaneous replication of RNA (without the help of enzymes) suggested that the error rate was around 1 in 20. That figure implies an upper limit on primitive genomes of about 20. Perhaps, with luck, the figure could be as high as 100. The bigger, the better. The reason is that the longer the RNA, the more opportunities there are to reduce the mutation rate, using RNA itself.

Single-stranded RNA often forms tangles where its component bases pair with themselves, notably in hairpin bends. The resulting complex shapes give it the ability to act as an enzyme—possibly to speed up chemical reactions that help correct errors. At the dawn of life, genetic replication might have been occurring very slowly, subject to high error rates, but perhaps primitive RNA enzymes emerged to help make replication more accurate by acting as a kind of template to arrange the chemical players in the game of replication. This is promising when it comes to creating a biochemistry that is complex enough to sustain life.

But, of course, we must remember Eigen's work on errors. Is it possible to make RNAs that are long enough to make error-correcting enzymes? Spontaneous RNA replication has an error rate of between 1 RNA spelling mistake in 20 to around 1 in 100. Let us assume the latter value (to be optimistic). The error-threshold suggests that the longest piece of RNA that can evolve is 100 bases long. Eigen says this is too short to encode for an RNA enzyme that might act as a replicase—that is, an enzyme that increases the rate of RNA replication and reduces the mutation rate.

How, then, can one make big enough chunks of RNA for the resulting RNA enzymes to guarantee accurate replication and vice versa? Some refer to this as Eigen's paradox: the error threshold concept limits the size of self-replicating molecules yet life requires much longer molecules to encode their genetic information. Eigen showed that without error correction enzymes, the maximum size of a replicating molecule is about 100 letters. But for a replicating molecule to encode error correction enzymes, it must be substantially larger than 100 letters. That is why the concept of an error threshold is crucially important in the context of the origins of life.

I focused on error threshold theory when I worked on my master's thesis with Peter Schuster in 1989. While he and Eigen had used infinitely large populations (although this sounds intimidating, they are actually easier to analyze), my task was to refine their work to cope with finite populations. The result was a neat little equation that I hoped would please some experts. It turned out that the maximum possible mutation rate was smaller in finite populations than in infinite populations. Therefore in the realistic setting of finite populations, the problem of the error threshold was even more acute. For a given mutation rate the maximum genome length compatible with adaptation (finding peaks on the fitness landscape and staying on them) was even shorter. That meant it was even harder to overcome Eigen's paradox.

Yet we know that this limitation was overcome. The planet is teeming with life that contains huge molecules. In our own bodies, enzymes are big, beautifully sculpted proteins designed with exquisite precision to speed up chemical reactions in a precise way and ensure, for example, that errors in replication are at a minimum. The presence of every living creature on the planet tells us that evolution has found a way to sidestep this difficulty.

HYPERCYCLES

It turns out that cooperation provides the solution to this apparent catch-22 of life. In this case cooperation comes in the form of what is called a hypercycle, a cycle of mutually dependent RNA molecules, each one undergoing chemical reactions to reproduce itself and, at the same time, help the next molecule in the cycle to reproduce.

This is a kinetic model, one that gives the broad brush picture of how the parts of the hypercycle work together. Each replicator is small enough not to run into the error limit and so can be made with high fidelity. Nor can any one replicator take over the brew: the ensemble is such that the RNA actors depend on one another for a successful performance. By establishing a hypercycle of individual RNA sequences, each one below the information error threshold, a larger genetic message could be stored, such as one capable of making proteins to check for errors and to repair them too.

How do hypercycles relate to the real world? Life on Earth today stores genetic information in the form of DNA. This is translated into RNA, which is used as a working copy of the information encoded in DNA and can also function as an enzyme, which in turn specifies the building blocks of bodies, proteins. In the RNA world, all these jobs were done by RNA, but one can extend the hypercycle idea to cover a rainbow of possibilities—RNA hypercycles, DNA-protein hypercycles, cycles turning within other cycles—but all of them share the same mathematical properties.

Hypercycles spin within the ecosystems that surround us, in the form of mutually dependent relationships between living things. Darwin expressed this idea in the following, elegant way: "It is interesting to contemplate an entangled bank, clothed in many plants of many kinds, with birds singing on the bushes, with various insects flitting about, and with worms crawling through the damp earth, and to reflect that these elaborately constructed forms, so different from each other, and dependent on each other in so complex a manner, have been produced by laws acting around us."

CHEATS

The scientist is not a person who gives the right answers, he's one who asks the right questions.

—Claude Lévi-Strauss, *Le cru et le cuit*

The hypercycle may solve the error problem but leaves a bigger issue, one first identified by John Maynard Smith. How did hypercycles evolve in the first place? Why should they remain stable if there are molecular parasites? Those who are unfamiliar with this work may scratch their heads, confused by the idea that molecules can be parasites, or baffled by how they can possibly "decide" to cooperate or to defect. This is an anthropomorphism. A molecule can be the right shape or chemical makeup to speed up a useful chemical reaction (a "cooperator") or could be the right constitution to disrupt it or divert cellular resources (a "defector") because it snips other crucial participants into pieces.

In the same way, the idea of a "selfish gene" should not imply that genes have actual motives, only that their effects can be described as if they do: the genes that get passed on to the next generation are the ones whose consequences serve their own interests, not necessarily those of the societies or even the organisms in which they find themselves. And when it comes to the molecular parasites of Maynard Smith, we are talking about something like an RNA that receives help to replicate but, being a defector, does not give any help in return. Once again, we are back to the Prisoner's Dilemma and the ever present tension between cooperation and defection.

We have already encountered Maynard Smith in chapter 5, on kin selection. This affable, owlish man was the father of evolutionary game theory and exerted an extraordinary influence on my field. He also did much to kindle my own enthusiasm and love of theoretical biology.

Maynard Smith began his career as an engineer and worked as a "stress man" during World War II, calculating the forces at work in airplane wings. This was an unforgiving way to hone his mathematics, since they made him fly with the pilot when they tested new designs, presumably so that he would naturally and swiftly learn from any of his blunders. He then studied biology at University College London because his muse, the great J. B. S. Haldane, was a professor there. Mathematics is traditionally a young person's game but Maynard Smith went on to write his first great theoretical paper at the tender age of fifty-three.

When I visited him in the 1990s at Sussex University, near Brighton in the south of England, he still crafted his computer programs in the quaint and ancient language of noncompiled Basic. Having never mastered a graphics program, he would plot his results on paper by hand—as he did before my own eyes. He would take me to the pub for a pint of bitter accompanied by fish and chips, that other great British contribution to world cuisine. Together, we would return home to "the white house" where I could stay the night; he would take me for walks on nearby verdant hills, sand dunes, and salt marshes. We would talk without end. I would listen in awe. He often prepared the ground for a new line of thinking by posing a carefully constructed question.

One of Maynard Smith's most important contributions was introducing biology to the game theory pioneered by John von Neumann. The story goes that Maynard Smith became stuck while plowing through a highly mathematical book on the subject. After a few pages, he decided to abandon the textbook and continue to develop his own thoughts. He came up with the concept of an "evolutionarily stable strategy": one that, once common, cannot be bettered by alternatives. He pondered what would happen if a population of players employing one strategy encountered an individual using another. The mutant strategy can invade and take over the population if the mutant can obtain a higher payoff than the typical member of the population.

In this way, Maynard Smith stumbled across a way to play games with a population, which was fundamental for thinking about evolution. What is remarkable is that if he had stuck to the textbook he would have found, a few pages later, the very similar concept of the Nash equilibrium, named after John Nash, whom I had encountered in Princeton. He might have been discouraged and never come up with the idea of an evolutionarily stable strategy. Perhaps Maynard Smith would not have become that famous if he had read on. Sometimes too much knowledge is a bad thing.

Maynard Smith made another great contribution when he warned that hypercycles, altruistic networks of molecular replicators, might have a major problem. He wanted to know how they could cope with parasites, free riders, and cheats. This could, for example, be a participant of the hypercycle that receives catalytic help but does not aid any other unit of the network. If such a parasite emerges, then the hypercyclic chain is broken. The presence of cheats that can exploit the hypercycle and divert its resources reminds us that though hypercycles solve the error problem, we now have to solve a deeper problem: How do these complex, interdependent communities evolve?

The answer should be familiar by now. We need a mechanism for the evolution of cooperation. Eigen had argued that cooperation might be possible if hypercycles were packaged in cells. If we assume that cells with the most successfully replicating—and thus cooperating—hypercycles divide most rapidly, then these cells would thrive relative to those that had hypercycles that are being challenged with cheats. By evoking a higher level of selection, acting between cells, hypercycles can shrug off parasites. This is a beautiful example of multilevel selection, where each cell acts as a group of replicators. Within cells, defectors might win. But cells without defectors can outcompete cells with defectors. Cooperation rules.

Maynard Smith also realized that during evolution there have been several major changes in the way genetic information is organized and transmitted from one generation to the next. He described the evolution over the eons from genes to chromosomes to cells to language in the book *The Major Transitions in Evolution*. It was fitting that Maynard Smith wrote this book with Eörs Szathmáry, a biologist who had come from Hungary like so many other great, influential scientists and mathematicians—such as John von Neumann or that most cooperative mathematician of all, Paul Erdős, who worked with hundreds of collaborators, on problems in combinatorics, graph theory, number theory, classical analysis, approximation theory, set theory, and probability theory. Indeed, PED also has a resident Hungarian, Tibor Antal, who has cracked some of our toughest problems. As the old joke goes, "If I have seen further, it is by standing on the shoulders of Hungarians."

Maynard Smith and Szathmáry charted how countless leaps and bounds in complexity and design are required to get from a formless void to modern life. Among the first steps, genes began to be linked together in packages known as chromosomes, which are essentially communities of interdependent genes. Chromosomes divide tidily during cell division to produce daughter cells containing identical sets of chromosomes and identical genetic material.

Thus chromosomes are communities of genes whose fortunes are intertwined. By linking with others to coordinate replication, each gene ensures that all the daughter cells acquire the full complement of cooperating genes. In this way, selfish interests promote cooperation. Our own cells contain tens of thousands of genes scattered across forty-six chromosomes, reflecting how an individual gene will flourish in a cell that contains all the others needed to sustain that cell. Indeed, some genes produce components that only make sense when stitched together with those made by other genes: there are many more different proteins in our bodies than genes as a result.

The relative advantage of having all genes linked on a chromosome dramatically increases with their number, so the more genes that cooperate this way the better. Once genes are linked together in a larger whole, then error correction machinery can evolve by natural selection. Linkage not only ensures that mutually complementary genes always find themselves in daughter cells but reduces the opportunities for defection because genes that are bundled together this way are replicated simultaneously: when one gene is replicated, they all are.

But, of course, there are still defectors to worry about in the form of selfish genes. For example, it may be that genes operating individually could replicate themselves much more quickly than those cooperating in a chromosome, which are obliged to wait their turn. Why should genes tolerate the straitjacket of the chromosome? Why not break loose? Why cooperate in the first place? Why not indeed? Well, of course, not all genes do cooperate for the greater good. One of the remarkable discoveries is that the entire human genetic code—genome—is swarming with selfish elements, the result of "intragenomic conflict."

Our inheritance has been molded and shaped by genetic parasites, selfish "duplicate me" instructions that have been passed down over the generations. With meaningless names such as Lines, Sines, Ltr retrotransposons, and DNA transposons, these parasites make up a significant fraction of our genetic code, representing around 13, 20, 8, and 3 percent respectively. Some, like Lines, code for protein machinery that inserts new Lines in our genetic recipe. Others, notably those called Alus, take advantage of the protein machinery produced by the Lines to reproduce. Indeed, selfish genes may well have been scrambling DNA since the dawn of life, some 4 billion years ago. In this way, natural selection has favored at least some genetic mavericks. But there are limits. A gene that is too selfish will end up killing its host and itself at the same time.

ALIEN THOUGHT

The discussion so far shows how cooperation is a key ingredient of evolution on our planet and it is logical that the same should be true for life everywhere. Given the vastness of the cosmos, the transition from prelife to life is likely to have occurred in many places and at many times. No wonder, then, that astrobiologists are hunting for evidence of alien life.

Some are looking for life far, far away, using powerful telescopes to hunt for telltale chemical signatures on Earth-like planets in distant solar systems. Others are looking much closer to home. Earth itself may even possess a "shadow biosphere" of alternative microbial life representing a second genesis, third genesis, and so on. In search of evidence, scientists are scouring ultra-dry deserts, drilling down into lakes buried under ice sheets, sending balloons into the upper atmosphere or other environments that are too extreme for life as we know it, and finding ways to detect the presence of alternative biochemistries.

Despite the enormous progress in understanding the events that may have occurred at the dawn of life, there is one aspect of Earth's life story that still amazes me. How did life get established so quickly? It could have taken as little as 200 million years to go from a lifeless planet to one where countless bacteria flourished. Now although 200 million years sounds like a long time, it is indecently rapid when compared to the "relatively simple" step of creating the first complex cells—Eukarya—from bacteria, which took 2 billion years. Going from "nothing" to bacteria seems to be a much more difficult task than going from bacteria to the more elaborate Eukarya.

We should not exclude the possibility that the first life was seeded from elsewhere. By this I don't mean that life arrived from a nearby solar system. A more likely bet is the following: the cloud of molecules, dust, and matter that agglomerated to form our solar system was laced with the leftovers of many dead star systems, some of which could have harbored life. Among this cosmic junk and detritus perhaps were the remains of earlier planets that carried bacterial spores. A rock impregnated with those spores could have planted the seed of all life on Earth. Even if the genomes of alien bacteria had been smashed by radiation, we know that some may well have been able to reconstruct themselves in water and start to replicate again.

We could be the spawn of molecules that won the first game of life far away, on a planet that once orbited another star in the dim and distant past. Astronomers are fond of telling us that we are stardust, a reference to how our bodies rely on the heavier elements that were once cooked up inside stars. But the time line that charts life's emergence suggests that we could just as well have been the result of cooperation between molecules that originated long ago on a rocky planet in an alien star system.

If life is a robust phenomenon—and I think it is—then it has originated often in our universe. Similarly, if intelligent life is a robust phenomenon it has also been generated frequently. Why have we not had a chat with ET yet? I think encounters between independent oases of intelligent life are rare because intelligent life is so unstable. Being smart is a fleeting phenomenon. It is self-destructive. Why? Because intelligent life often fails to solve the biggest problem of all: the problem of cooperation.

CHAPTER 7

Society of Cells

We are all cells in the same body of humanity.
—Peace Pilgrim (Mildred Lisette Norman)

They're among the most important fossils of all, mysterious monuments left by the very first organisms on Earth. Called stromatolites, some are smaller than a finger and others larger than a house. Among the most remarkable can be found in the Pilbara region of Western Australia, in the baking heat of what is jokingly called the North Pole Dome. There, white, red, and black rocks are dotted about in the grass. These stromatolites are thought to be little changed from when they first formed around communities of microbes an astonishing 3.43 billion years ago.

Some look like domes or upside-down ice cream cones. Others are small, conical shapes arranged a little like an egg carton. A few are crest-shaped, or even bear a passing resemblance to the ears of Mickey Mouse. Studies of their microbial descendants, known as cyanobacteria, suggest that these peculiar rocks slowly formed when mats of microorganisms trapped sediments and precipitated carbonates. Because of their great and extraordinary old age, these stromatolites represent the first chapter in the rise of cooperation, a vestige of the ancient microbial communities that were among the first living things on Earth.

They are singular monuments. But as evolution shapes new kinds of cooperation, and cooperation heralds ever more inventive ways to construct an organism, novel ways to destroy and to defect also emerge. The stromatolites are a testament to cooperation among our ancient single-celled heritage, when they began to work together, and warn of the powerful and primitive forces that can be unleashed when that ancestry starts to reassert itself. I have come to think of these striking rocks as both milestones of the rise of living cooperation and as tombstones that warn of its potential fall.

Such single-celled creatures seem distant relations of modern life, yet in every sense, they are very much with us today. Their descendants are ubiquitous, clever, and unstoppable. They are able to eke out an existence in brutal conditions, from a subzero chill to the hell of scalding, acidic hot pools. They bask in the extreme saltiness of the Dead Sea and enjoy the caustic delights of soda lakes. They can be found in bone-dry deserts, such as Atacama. They thrive in the crushing depths of the ocean where temperatures are well in excess of 100 degrees Celsius. They lurk in the muds deep below the sea floor, along with toxic and radioactive sludges. Along the way, bacteria have invented all of biochemistry. And, at a stroke, they invented much of the machinery that turns in our bodies.

They also evolved ways to cooperate. Multicellular strings of bacteria were born around 3.5 billion years ago. Filamentous bacteria—so named because they form chains—kill themselves to yield precious nitrogen for the good of their sisters. Every tenth or so cell commits suicide for the benefit of this communal thread of bacterial life.

Another, quite different kind of microbial cooperation was revealed by the dogged efforts of Lynn Margulis at the University of Massachusetts, Amherst. She proposed that more complex "higher" cells were the result of symbiosis, when single-celled creatures became so closely associated that they worked as one. For example, around 1.8 billion years ago, there was an important moment when one kind of wriggling bacterium invaded another. Perhaps the former was seeking food. But this particular parasitic infestation suited both parties and evolved so that the participants formed a long, harmonious, and fruitful truce. This is what Margulis calls symbiogenesis and it led to the formation of higher cells, known as the Eukarya.

Thanks to this cooperation, a new, more complex kind of cell appeared on Earth. Whereas bacterial cells are relatively simple and known as prokaryotes, the newer cellular consortia, known as the eukaryotes, are the building blocks of plant and animal bodies, including our own. These cells contain organelles—which divide up the task of cellular life as organs do for a body—including a nucleus where their DNA resides. These organelles are the leftovers of earlier episodes of microbial mergers and acquisitions.

Peek inside your own cells and you will find these Russian doll coalitions. The most obvious example of symbiogenesis comes in the form of little lozenge-shaped structures called mitochondria. Not only do they look like bugs, they even have their own separate DNA that is passed down the maternal line, from mother to child. Our cells are driven by these descendants of bacteria that hundreds of millions of years ago traded chemical energy for a comfortable home. These organelles now power our muscles, our digestion, and our brains.

Nature has mixed and matched simpler creatures this way for eons, underlining once again how cooperation goes hand in hand with construction. The plants in the window box, trees in the garden, and the broccoli at the greengrocer's all date back to ancient ancestors that became verdant only around 2 billion years ago, when they teamed up with smaller green creatures that could catch sunbeams and turn them into food. Indeed, researchers even discovered a tiny ocean creature on a beach—Hatena ("mysterious" in Japanese)—that seemed to be engaged in the very process of going green.

Then came a second wave of cooperation when these more complex cells themselves teamed up to form communities. The result—a multicellular organism such as a dog, cat, or you and me—profits handsomely from a division of labor between their component cells. More than 600 million years ago ctenophores—common and fragile jellies with well-developed tissues—were probably the first off the blocks when multicellular life began to diversify. Sponges mark another early example of solitary cells that came to cooperate in more complex bodies. They have different kinds of cells—digestive cells, cells that secrete the spicules (segments of the body skeleton) of spongy proteinaceous material, and so on—which can communicate with one another and divide up the labor of life to work together as a single individual. The draft genome sequence of one, the Great Barrier Reef demosponge (*Amphimedon queenslandica*), reveals genetic mechanisms that allowed individual cells to work together, from those that help them adhere to one another to those that suppress individual cells that multiply at the expense of the collective.

In fact, coagulation of complex cells into cooperative communities was such a winning strategy that it evolved several times. Animals, land plants, fungi, and algae all joined in a communal life, and not just of their own kind. Coral reefs, the biggest living structures on Earth, are formed by an enduring partnership between an animal (polyp) and plant (algae), held in a permanent embrace by a skeleton of limestone. Another marine example is the Portuguese man-of-war, which can measure more than 150 feet from its air bladder to the tips of its tentacles. Many think of it as a jellyfish, but in fact it is a siphonophore—a colony of minute individuals.

Given how many times multicellularity evolved, it seems unlikely that there is a single explanation for its origins, save that the same basic strategy—cooperation—was the right answer when it came to dealing with various problems. Even though cells within a community still reproduce independently, huddling together in a community may represent the first step in the transition to multicellularity if it offers some kinds of benefit to the group. The incentive for cells to get along could have been that it helped to shrug off defectors, such as parasites. Or it might have helped to develop a better way to move about and exploit the available sources of food and energy. Or perhaps it allowed cells to mount a more effective form of defense.

The next step leading to the evolution of linked reproduction could have been the exchange of resources between cells that belonged to the same community. We can get insights into why this occurred from strange creatures called slime molds, which represent a fascinating halfway house between single-celled and multicellular creatures. A colleague at Princeton, John Bonner, used to enthrall me with his accounts of the peculiar life cycle of these so-called social amoebae, which are common soil organisms. They graze their separate ways as individual amoebae and, when the going gets tough, club together to form a multicellular organism that has a remarkable ability to move and orient itself, even send out spores from atop a stalk created of cells that are willing to sacrifice themselves for the greater good.

Beyond the slime molds lies the evolution of full cell differentiation, where some cells do one job in the serious business of life, while other cells do something different. When it comes to building a human body, for example, the organism starts out with "blank-slate cells" in the early embryo, called embryonic stem cells, which then multiply and branch out into any cell type. In all there are more than two hundred types of cell—brain, heart, muscle, skin, and so on.

Multicellularity gave the world so much. Just look at all the incredible creatures that inhabit our planet, from butterflies to whales and millions more besides. But, like all good things, this teeming diversity comes at a price. If the Prisoner's Dilemma teaches us one lesson, it is that every stage of cooperation comes with the risk of defection. Cooperation is never stable. There is a surge in cooperation during early development and childhood, but our cells begin to rebel with age.

I want to focus on one example of cellular defection with which everyone is familiar. We are built of units that can replicate on their own, that can revert to their own originally selfish program and start to act like any self-respecting microbe. This teaches us that we can't ignore the central lesson of the Dilemma. There are cells that will turn against their host, even though their long-term future depends on being part of a multicellular organism.

One of these defectors is the cancer cell, which turns against the interests of the body. When these errant cells discover how to undermine cooperation, death is the ultimate prize. Around 7 million people perished of cancer in 2004—representing 13 percent of worldwide deaths. The toll is projected to continue rising, with an estimated 12 million cancer deaths in 2030. This awful disease is but one consequence of a collapse of cooperation, when our single-celled heritage reasserts itself. Cancer is the price that we pay for having complex bodies built by an extraordinary level of cooperation.

WHEN COOPERATION FAILS

Divided we fall.
—Matthew Walker

Everyday acts, from eating to drinking to talking, rely on exquisite coordination among a vast range and number of our cells, from those within the eye to muscle cells in the jaws to the nerves that pass electrical impulses around the brain. To maintain this concerted activity, the body must be able to grow and repair itself. Our organs rely on a constant turnover of cells. Every day we need several hundred billion new blood cells. The outer layers of the skin also need to be replaced constantly, so this, the largest organ in the body, is in a constant state of repair. The same is true for the lining of the lungs or the gut or the milk ducts in the female breast. When cells in these structures do break ranks, there is a breakdown of cooperation on the cellular level.

Cancerous rebellions can arise when cells get mutations that alter their program. Now, when I say mutation, I mean any change to the DNA inside a cell that occurs for any reason, whether a spelling mistake during cell division, missing sections of DNA, DNA spliced in by a virus, or muddled chromosomes. Most mutations don't do anything, but some mutations can be dangerous: they can induce the cell to divide when it should not. Or a mutation might prevent the cell from committing suicide when the body tells it to do so for the greater good. A mutation might block a cell from taking a break to repair itself. If left unchecked, cells can proliferate to fulfill their own selfish agenda, not that of the organism. Cancer is a disease where individual motives return to dominate.

Traditional biology focuses on the evolution of organisms in an ecosystem and we can now focus on how the same process creates cancer. Remarkable examples of evolution are to be found in the oncology clinic, where cells—tumor cells—mutate and change in an extraordinarily complex environment, that of the human body. Within our tissues and organs, cells in tumors face diverse selection pressures which favor mutant cells that are better able to survive, for instance those that divide faster and those that are less likely to perish.

Cancer starts with the origin of a single cheat: this solitary cell has a dangerous mutation and multiplies to form a small lesion. Now a few thousand cells have this mutation. Nothing might happen for years. Or nothing might ever happen at all. If another mutation allows this ensemble of cheats to grow, it could give rise to what is called an adenoma. Adenomas can become large and contain perhaps 100 billion cells. But they are still constrained and encapsulated by surrounding tissue.

All the while the cheats can accumulate several more mutations. Many are deleterious and make the cheats worse off. Even more are neutral and do not alter the cell's behavior. But a small fraction of these changes enable the cells to grow with more vigor. Eventually there might be a mutation that allows them to become "locally invasive." This is the transition from adenoma to carcinoma. The carcinoma grows and invades the surrounding tissue, but the tumor still remains confined to one part of the body for some time. It could be removed by surgery. Eventually, if the tumor has not been successfully eradicated, cells will arise that have the ability to travel elsewhere in the body. This is the most deadly aspect of cancer, called metastasis.

In metastasis cells set up homes in distant sites around the body. For example, colon cancer tends to generate liver metastasis. A surgeon can no longer remove all of the secondary tumors. The remaining hope is to use chemotherapy. Traditional forms of this treatment rely on the fact that, because they tend to divide faster than ordinary cells, cancer cells are more likely to succumb to poison. But because chemotherapy is not very specific, bluntly attacking fast-dividing cells whether cancerous or not, there is always collateral damage: side effects, such as nausea, baldness, and deafness.

We do have natural anticancer mechanisms. But they are far from perfect. Evolution works hardest to keep us alive to reproduce. In fact, reproduction is the major game in town. Once a person has successfully passed his or her genes on to the next generation, those genes "care" less if that creature lives on. In this way, genes are favored that shift the balance toward investing scarce resources in mechanisms that promote reproductive fitness and the maintenance of egg and sperm cells—the germ line—rather than the body, or soma. This echoes an ancient idea: Epictetus the Stoic philosopher once wrote that if we were useful alive, "should we not be still more useful to mankind by dying when we ought, as we ought?"

There is strong selection pressure to ensure that cancer does not occur too early, rather than to build us to last forever. Natural selection is relatively indifferent to the fate of the old, since their relevance to evolution is waning fast (unless, for example, elders use their wisdom to help their families to cooperate and to survive). Natural selection is honed by the death of the young, who have not passed on their genes, and by those who survive to bestow their genes on their children.

The opportunity to sire the next generation was particularly narrow for our ancestors. For most of human evolutionary history, people did not live as long as they do now. Genetic variants that could help to keep us cancer free in our eighties made no difference to the lives of our ancestors, who were likely to perish relatively young from starvation, infectious disease, an attack by a wild animal, or a sharp blow from the flint axe of a rival.

That is one reason why, in today's greying society, cancer is a growing problem. There are various possible mechanisms by which cancer becomes more prevalent with age. Imagine a gene that prevents cancer in children but increases the risk of cancer in the elderly. Even more likely: imagine a gene that enhances the ability to have more children at the cost of a higher risk of cancer later in life. Nature tunes female anatomy to increase her capacity for conceiving or nourishing children and, in return, cuts the probability of living an extremely long life.

Cancer is also linked with lifestyle, and that has changed radically over human evolution. One example is exposure to pollutants and carcinogens. Thanks to the efforts of the Oxford statistician Sir Richard Doll, along with many others, smoking is the best-known risk factor, being linked with lung and other cancers. Some chemicals in tobacco smoke can directly damage part of our DNA, including key genes that protect us against cancer. Others interfere with our bodies' defense systems and prevent them from repairing damaged DNA, making it more likely that damaged cells will eventually turn cancerous.

After smoking, obesity is the lifestyle factor that presents the highest preventable cancer risk. Our bodies are not adapted to deal with a constant oversupply of energy. Yet thanks to the rise of industrial farming we can eat whenever we want, and to excess as well. Activity levels have also declined. There is now an epidemic of obesity in the West. So there is a deep link between the way our bodies have evolved and the rise of cancer. The resulting breakdown of cooperation is something that I have tried to explore with mathematics.

My hope is eventually to help doctors form a quantitative understanding of the process of cancer formation and growth that can guide the way they treat patients. In fact, more than this, I would hope that we can make treatment as predictable as engineering: only this time we are designing the body to encourage cooperation and to resist rogues, breakaway groups, and cellular independence movements. I would like to see the day when we can take the genetic makeup of a cancer in a particular patient, then customize the treatment to clobber the machinery of the cancer cell and—without causing collateral damage to normal cells—restore cooperation.

MUTATIONS

My interest in cancer first became an obsession when I was invited to Rockefeller University in New York City by its then president, Arnold Levine. He wanted me to find a way to portray cancer in mathematical terms. Levine was famous for his simultaneous discovery (with David Lane of the University of Dundee) in 1979 of a cancer "causing" gene called p53. At first p53 was believed to be an oncogene—one that accelerated the cell cycle, so it minted more cells—but genetic and functional data obtained a decade after its discovery showed that this was not right.

In 1989 Bert Vogelstein, working at Johns Hopkins University in Baltimore, did important spade work to show what p53 does in the body to make it play such a central role in cancer. Genes are the recipes to make proteins, and the one made by this gene is a tumor suppressor. Vogelstein found that it does everything it can to prevent cancer. But if it is mutated so that it can no longer suppress tumors, cells can go wild. Damage to the genome is no longer repaired and/or the damaged cells no longer undergo apoptosis, or programmed cell death (when they commit suicide). Mutations that harm the workings of p53 are very dangerous.

Vogelstein and others found that this gene is mutated (that is, inactivated) in about half of all human cancers. The gene, nicknamed the "guardian of the genome," is responsible for making a protein that lies at the heart of a control system that monitors genetic damage. A little disturbance, and the cell pauses to fix itself. Too much disturbance, and apoptosis kicks in. But when the gene, and thus the protein, is faulty, the cell can keep dividing even when it's genetically damaged. Such is its significance in the development of cancer that in 1993 the journal *Science* even dubbed the protein made by p53 the "molecule of the year."

I became fascinated by Vogelstein's work and decided to drop everything I was doing at that time to study cancer. I wanted to understand this killer in terms of the one element of biology that made perfect sense to me: evolution. I reasoned that if I could provide a Darwinian insight into what turns an ordinary cell that works with the body into a mutant that works against it, then I could calculate something useful about cancer, rather than speculate about why that society of cells we call a human being sees more tumors in old age.

I sent Bert an email and asked if I could visit. The next day, an email pinged back from Baltimore: "I am happy to see you and hope that you can do for cancer what you did for language." I was surprised that he knew of my research on the evolution of language (see chapter 9). I was even more surprised when, a little later, someone told me that "everyone wants to see Vogelstein" and that he was usually much too busy to oblige. This was underlined when I found out that his research was the most cited on the planet. Whenever academics quote from or paraphrase a study or a paper written by one of their peers, they use citations to acknowledge that they have done so. Vogelstein is the most cited of all scientists on Earth, dead or alive. He is an inspiration.

Eventually, I found myself outlining my cancer ideas before his group in a seminar room in Johns Hopkins. The man sat before me, a crouched figure wearing a baseball cap. I could not help but notice there were electric guitars hanging up on the wall. It turns out that members of his lab play in a rock band that performs regularly in night spots dotted around Baltimore. Vogelstein himself is a keyboard player. The name of their band is Wild Type (appropriately for a geneticist, where this refers to an organism, strain, gene, or characteristic as it occurs in nature; in other words, the wild type is the opposite of a mutant) and was born when a few of the lab members started fooling around with guitars in the conference room. The results were so awful that the methadone clinic next door complained that the racket was upsetting the recovering drug addicts.

There was something about Vogelstein's demeanor and sharp mind that seemed familiar. I decided that he must be the Bob May of cancer. My suspicion was confirmed after I cracked a joke, one I have told many times to sum up what biologists think of the efforts of mathematicians. A stranger approaches a shepherd and his flock and asks, "If I tell you precisely how many sheep you have, can I have one?" The shepherd agrees and, after a single look, the man declares without a moment of hesitation, "Eighty-three." He picks up an animal and turns to leave, at which point the shepherd offers a return challenge: "If I guess your profession, can I have the animal back?" "Sure." The shepherd says, "You must be a mathematical biologist." The stranger was amazed. "How did you know?" "Because you picked up my dog." After the talk, Vogelstein ushered me into his office. "I don't know much about mathematical equations but I can help prevent you picking up dogs."

Because of his pioneering studies, some of the best-understood tumors of all are to be found in the colon, the last lap of the digestive system. Bert spent a good ten minutes rummaging around for a sample that he was keen to show me. Eventually he handed me a microscope slide. This showed the first stage of colon cancer, called a dysplastic crypt. After taking just one look, I realized that I had to change my model of cancer formation in an important and fascinating way. Quickly, and hoping that no one would notice, I returned my dog to his fellow sheep.

In my original model, I had studied the emergence of colon cancer in a large population of identical cells. As soon as I looked at Vogelstein's slide, I realized that the cells weren't identical. In fact, there was an intricate geometric arrangement of cells in the colon, and I would need to formulate a model that really took this arrangement into account. From my previous work (for instance in my studies of the primordial pizza) I knew only too well that evolutionary dynamics tend to behave very differently in structured as opposed to unstructured populations.

The slide showed that the cells in the tissue lining of the colon are organized in "crypts," each of which is shaped like a tiny ice cream cone. On the slide, I could see that one crypt that had become cancerous was spreading toward its neighbors. I realized that colon cancer had to begin in a crypt, itself a small population of several thousand cells. So my model first had to deal with a crypt, and then work out what followed in a population of crypts. In short, I had to work out the probability that one cell in a crypt would become mutated and then the probability that the progeny of that cell would take over the crypt.

The clever design of crypts is a direct consequence of how tissues are renewed by stem cells, the "parent cells" in the body. This is particularly important because, given how much material we digest, the lining of the colon is forever being replenished. A few stem cells sit at the base of the crypt and divide once per week to produce colon tissue cells. As these cells become more differentiated and move up the crypt, their progeny divide more and more quickly. By the time they reach the top of the crypt they are dividing around once per day, until they eventually die at the hands of apoptosis, a self-destruct program built into every cell.

My later work showed how this cellular structure, with a few slowly dividing stem cells at the base of a pocket, and the fastest dividing cells at its top and closest to death, puts a brake on evolution. This is precisely why the colon is populated with crypts. Mother Nature has honed their design to provide fewer opportunities for tumors to evolve.

You can actually prove that these structures, which are seen in tissues with high turnover, have optimal anticancer properties. By washing out the mutations from the fastest-dividing cells as quickly as possible, cancer is delayed. That leaves only the rare and relatively slowly dividing stem cells to be really vulnerable to cancer causing mutations.

There were wider implications of this work on that unwanted form of evolution that we know as cancer. The evolution of crypts was the way that humans, and presumably our ancestors, evolved to oppose the progression of cancer cells. Evolution consists of mutation and selection. The clever tissue design of crypts suppresses selection. In this way the arrangement of the players (the stem cell and its descendants) can alter the pace of evolution, in this case making it harder for cancer to develop. Therefore, the tissue architecture is designed to maintain cooperation among cells in the body. So I began to think in general about how the relationship between the different players in a game—the topography—can affect evolution. Spurred on by these considerations, I would come up with the idea that paved the way to evolutionary graph theory, which I will return to in chapter 12.

Around the same time, I began to make progress on the study of a cancer that seemed to be triggered by a single genetic event, a blood cancer called chronic myeloid leukemia, CML. Bert Vogelstein suggested that I work with Charles Sawyers of the University of California, Los Angeles, a leading light in the effort to understand the crafty survival strategies of cancer cells and the world's expert on CML. In turn, Charles told me that his colleague Tim Hughes, at the University of Adelaide in Australia, would have the perfect data that were needed for our analysis. Altogether it took three years to assemble the data and perform the analysis together with Yoh Iwasa and Franziska Michor, an inspired theoretical biologist who is now at the Harvard School of Public Health.

The good news for CML treatment is that recent years have seen the first molecular targeted therapy that's aimed at cancer cells: imatinib, or Gleevec. But the best anticancer drug yet manufactured does not seem to target the causative cell population: the cancer stem cell. Gleevec achieves its great success by preventing differentiated cancer cells from growing, not stem cells. Moreover, in some patients the leukemia eventually develops a resistance to Gleevec. In the wake of these mutations, a resistant subpopulation of leukemia cells can grow despite the presence of Gleevec. In the light of this understanding, new drugs are now being developed to oppose these resistant mutations.

In a joint project with Tibor Antal and Ivana Bozic, a mathematician from Serbia, I am now collaborating with Bert Vogelstein to understand how mutations arise in growing populations of cancer cells to drive the overall progression of disease. We are studying the growth of adenomas, carcinomas, and metastasis, using a mathematical model where every new driver mutation enhances the growth rate of the tumor.

The first driver mutation gives rise to the initial expansion of the tumor. However, the deadly aftermath very much depends on when the second driver mutation occurs. In some of our digital "patients" the second mutation arises within a year or two. These virtual patients have large tumors and many mutations within a decade after the very first mutation. Theirs is a fast and deadly disease progression. But in other patients the second mutation can wait eight years to arise. In this case the tumor is relatively small for ten years or more. Fatalities are down to a throw of the evolutionary die. We all rely on the continuing cooperation of our cells. The struggle to cooperate in our body is a matter of life and death.

HOW TO RESTORE COOPERATION

In order to develop new ways to combat cancer, we can learn much from nature herself. The most basic instinct of any cell in our body is to divide, so over millions of years evolution has already come up with some clever mechanisms to curb this selfish instinct and make our cells resist the deadly mayhem of cancer. There are genes that tirelessly work to keep the cell's genetic material, or genome, free of errors. Others ensure that cells divide cleanly. Most cells are continuously listening to signals from their neighbors, soothing them when they are doing OK. If they do not get these chemical murmurs of reassurance, they might kill themselves— the cell adopts Plan B: apoptosis. So, for example, if a liver cell enters the bloodstream and lodges elsewhere in the body, it gets the wrong signals and self-destructs. You can think of the body as a hive and these signals as the equivalent of peer pressure to conform and to do the right thing.

There are other ways that we are protected against the development of tumors. There is evidence of "immune surveillance," where early cancers are detected and killed by the immune system. Special white blood cells, called T cells, circulate through tissues where they recognize and destroy cancer cells. Gradually, however, tumor cells get selected by a Darwinian process of survival of the fittest, so that they can begin to ignore T cells, then eventually shrug them off.

Evolution, in her wisdom, has come up with another way to avoid or delay the onset of cancer, as we have already seen. She has constructed us out of hierarchies of cells. The rulers of these hierarchies are slowly dividing, long-lived parent cells that we call stem cells. These, the cellular equivalent of royalty, spawn lineages of more differentiated cells that divide more quickly and are short-lived. For example, the blood cells in our bodies—red cells, white cells, and platelets—start out as hematopoietic (blood-forming) stem cells in bone marrow, the spongy inner part of our bones. An organ such as the skin calls upon relatively few skin-specific stem cells to produce intermediate cells that in turn produce skin cells, which themselves cannot reproduce. We can thank this design for our long lives. And they are long, when measured out in the fundamental time span of our basic units. Typically, it takes a cell in our body a few days to divide.

Such insights might help doctors make sense of a diagnosis and direct the development of new treatments, such as immunotherapy, where a patient's anticancer immune response is made more effective. Through fundamental understanding, evolutionary biology can help patients to be winners in the battle against defecting cells. Just as the wings of insects, birds, pterosaurs, and bats each evolved independently to tackle flight and not from a common winged ancestor, so the way that cancer cells emerge and come to shrug off new drugs can reveal common reasons why cellular cooperation breaks down. By understanding these discordant cellular overtures, we can understand what makes a cell turn selfish and cancerous.

Ultimately, the cure of cancer will rely on advances in our understanding of our environment so that it can be designed to discourage the breakdown of cellular cooperation, and on the use of leading-edge therapies such as genetic engineering to restore cooperation or even to build additional control mechanisms into our cells to make sure that they are unable to pursue a selfish agenda.

The French poet, painter, and filmmaker Jean Cocteau once remarked: "You've never seen death? Look in the mirror every day and you will see it like bees working in a glass hive." Well, I have reflected on the deadly effects of cancer in that hive of cells that we all know as the human body. I have modeled how those cellular bees give up cooperation to work against the body by rekindling something of their ancient single-celled origins. I hope that by taking an evolutionary view of why tumors grow and spread, doctors can devise novel cancer treatments that make cells work in harmony once again. The threat of cancer would greatly diminish if we could prolong the duration of healthy cooperation among the cells in our bodies.

CHAPTER 8

The Lord of the Ants

A child comes to the edge of deep water with a mind prepared for wonder.
—E. O. Wilson

She's the czarina of cooperation. The sole sovereign of a society of millions. The most powerful potentate of pulling together. Measuring an inch or so long, the leaf-cutter queen lies at the heart of her sprawling subterranean empire. These dark red *Atta* ants abound in the New World tropics. Nests can be gigantic, stretching underground for up to eight meters in a network of tunnels, ducts, and chambers that extends over an area of some fifty square meters. All around lie tens of tons of soil that have been excavated by the queen's much smaller, mostly female, subjects. During her reign of up to fifteen years, her 3 million short-lived inhabitants divide up dozens of jobs to work together as one. Hail Atta, queen of cooperation.

Each megalopolis of these social insects, coordinated by a complex chemical language, is greater than the sum of its parts, creating new levels of organization among its seven physical subcastes. The castes vary some 200-fold in terms of weight and eightfold in terms of head width. In all, the seven kinds of ant carry out a total of around thirty tasks.

One caste cuts foliage and leaves—their mandibular muscles make up one-quarter of their entire body mass—and some tropical ecologists estimate that the leaf-cutter colonies may harvest up to 17 percent of the total leaf production of a tropical rainforest where they thrive, in

Mexico and Central and South America. In as little as a day, a nest's denizens move back and forth along well-trodden foraging paths to defoliate a single tree. In a year, some species are able to harvest up to 470 kilograms of dry plant biomass.

Another caste hauls the leaf fragments back to the nest using horizontal foraging tunnels, superhighways that can extend up to six meters or more. A third dices the leaf up still further in this assembly line operation. However, the ants do not themselves eat the leaves they cut. By applying fecal droplets enriched with digestive enzymes, they turn the finely diced leaves into compost to grow fungus. Workers pluck nutrient-rich swellings known as gongylidia that form on the thread-like fungus and feed them to the colony's larvae. They are peaceful mushroom farmers.

Remarkably, the colonies cannot survive without their *Leucocoprinus* fungus and the fungus is found nowhere but in these colonies. They do not do farm the precious fungus on the surface. Instead, they grow fungus in underground chambers that can reach the size of a football. In all, a single leaf-cutter nest may harbor a thousand such chambers. The smallest ants tend the fungus gardens and use antibiotic-producing bacteria to ensure their crop remains free from disease. They weed the gardens too, removing competing fungal strains, and keep it at an ideal, slightly acidic, pH. Thus the colony farms depend on three-way cooperation between ants, bacteria, and fungi.

Thanks to the digestive powers of the fungus, the ant larvae are able to consume the otherwise unpalatable harvest of tropical forests whose leaves are laden with chemicals designed to deter browsers, such as terpenoids and alkaloids. These ants are able to grow a monoculture year after year without disaster, and they use their antibiotic so prudently that there's no sign of the antibiotic resistance that now plagues human medicine.

As a rule, riskier jobs are left to older workers who are destined soon to perish. Examples include waste disposal and defense. If the colony is disturbed, soldiers storm out of the nest and attempt to overpower the aggressor. While we send young men to war, ants send their old ladies. Soldiers are a caste of elderly females, each with a three millimeter wide head and well-developed mandibles. Their bite can penetrate human skin. So tenacious are their jaws that indigenous people in the Americas use them as sutures for holding the edges of wounds together.

Ant wonders do not end there. To establish new colonies, young males and females depart on a mating flight each year. A winged female ant mates with up to eight males, typically from other colonies, high in the air during a nuptial flight and stores all the accumulated sperm for the rest of her life. After the flight, all males die. The young queen digs a vertical shaft and creates a chamber at the bottom, which serves as the first of her own nests. There she deposits a fungus wad from her original colony (carried in a pouch on her body, called the infrabuccal pocket) to start a new fungus garden, the success of which is crucial for the future of the new colony. She removes her four wings, eats them, and lays her first handful of eggs. When the first workers emerge, they begin to eat, and to tend, the fungus culture. Groomed and fed by her workers, she can lay 20 eggs per minute, 28,800 per day, and 10.5 million each year. During her lifetime, a queen can produce more than 150 million daughters.

Ant colonies have much to teach us about the secrets of cooperation and advanced social behavior. They are one of the most successful forms of life, with at least 14,000 species. They have perfected ways to divide up labor to cooperate that appears more collegial than anything we do. They developed agriculture and architecture millions of years before our ancestors had even managed to walk upright. They are able to wage war. Unlike most ant species, army ants do not construct permanent nests but forage incessantly, pouring into an *Atta* nest and looting it if its citizens cannot mount an adequate defense against the marauders. Ants are also able to cooperate with other species, so that their lives are knotted together in a ruthless yet highly successful struggle for survival. For example, some species tranquilize aphids with drugs to keep them docile and "milk" them with their antennae for a treat of sugary honeydew.

In these eusocial societies, some group members surrender part or all of their personal genetic fitness to benefit fellow members other than their own direct descendants. It is the most advanced form of cooperation to be found in insects. And it works. Social insects are the most abundant of the land dwelling arthropods. Ants are perhaps the premier example, with the global mass of ants (up to 10 million billion of them, give or take) being roughly the same as the global mass of people. Even more impressive, these societies have thrived since the days of the dinosaurs. They offer an extraordinary glimpse of how cooperation can emerge from competition.

RISE OF THE SUPERORGANISM

A well-flavoured vegetable is cooked, and the individual is destroyed; but the horticulturist sows seeds of the same stock, and confidently expects to get nearly the same variety. . . . Thus I believe it has been with social insects: a slight modification of structure, or instinct, correlated with the sterile condition of certain members of the community, has been advantageous to the community: consequently the fertile males and females of the same community flourished, and transmitted to their fertile offspring a tendency to produce sterile members having the same modification.
—Charles Darwin, *On the Origin of Species*

Atta is not alone when it comes to being a master of cooperation. Think of a single worker bee. This solitary insect is as useless as a severed finger. But when attached to a colony, the bee becomes as useful as a digit on a hand. Now that bee can probe for the nectar of flowers and, once it has found a rich new source, point fellow hive members to these rich feeding grounds. Rather than gesture with wing or antenna, it uses a dance rich with symbolic information. In the same way that many factors and proteins coordinate the activity of cells in that organism known as the human body, so dozens of chemicals made by honeybee queens, workers, and brood play a role in social organization. Beehives are organized around an egg-laying queen tended by workers who, during their lifetime, make the transition from hivebound duties, such as nursing larvae, to more far-reaching jobs, such as foraging for food or defending the nest.

The different cell types in multicellular organisms are analogous to the different castes in a beehive, with workers constituting the soma—body tissue—and the queens representing the germ line, eggs and sperm. And, just as the body has mechanisms to weed out sickly cells, by apoptosis, a bee colony can regulate the lifespan of its members. The genome of our bodies is "optimized" by natural selection to build a good level of cooperation between germ line cells and soma cells with the help of apoptosis and various other processes. The same goes for breeding "good" workers and "good" queens. When I say good, I mean that they successfully reproduce and cooperate.

But, as we have seen in earlier chapters, there is a dark side to cooperation that comes in the form of parasites, cheats, defectors, and other lowlifes. In a healthy hive, workers identify and terminate cheats and abnormal colony members, ranging from embryos to adults. So long as this regulation continues, the colony thrives. However, if the types of worker that enforce order become too few, or if hive members change into malignant forms that can sidestep control mechanisms to replicate aberrantly, order is replaced by anarchy that ultimately leads to the decline and fall of the bee society.

There are some well-documented examples of the chaotic collapse of bee society. Take the relocation of the Cape honeybee by beekeepers from southern to northern South Africa in 1990. The result was the widespread death of managed African honeybees, *Apis mellifera scutellata*. The unhappy episode revealed the way in which insect society is susceptible to exploitation by rogue workers. In the African honeybees, these rogues began to harness the brood-rearing capacity of the colony to enhance their own personal reproductive success. It turns out that the billions of *A. m. capensis* workers now parasitizing South African honeybee colonies are all descendants of a single worker that was buzzing around during 1990. Their explosive growth has been likened to a social cancer.

Because of these many parallels between multicellular organisms and multicreature societies, from the division of labor to cancer, ant nests and beehives are known as "superorganisms." The term (Latin *super* = above; Greek *organon* = tool) was coined in 1911 by the great American ant expert and biologist William Morton Wheeler (1865–1937) in an essay titled "The Ant-Colony as an Organism" and is defined as "a collection of single creatures that together possess the functional organization implicit in the formal definition of organism." Wheeler was told, on receiving an honorary degree from Harvard, that his study of insects had shown how they, "like human beings, can create civilizations without the use of reason."

But there is a puzzling feature of these societies that has been largely overlooked by investigators: the phylogenetic rarity of eusociality. By this I mean that of the 2,600 or so living taxonomic families of insects and other arthropods currently recognized, only fifteen are known to contain eusocial species. When it comes to vertebrate (backboned) creatures other than humans, only one—the naked mole rat—has achieved the same grade of social organization. Why is eusociality so rare? The mystery is deepened by the knowledge that once eusociality takes off, it is amazingly successful. The living mass of ants alone comprises more than half that of all insects and exceeds that of all terrestrial nonhuman vertebrates combined. The answer to this riddle lies in understanding how cooperation led to the emergence of superorganisms.

ANT WONDERS

The achievements of colonies of leaf-cutter ants have been hailed as "one of the major breakthroughs in animal evolution" by Edward Wilson at Harvard University. He should know. Wilson has studied ants for more than fifty years. His entire career seems to have been shaped by the sagacity of King Solomon, who remarked in Proverbs, "Go to the ant, thou sluggard; consider her ways, and be wise." Wilson has invested much thought in the origins of eusocial species. Wilson, Corina Tarnita, and I have worked together on what he calls a "climactic project" that aims to explain the origin of eusociality using the mathematics of cooperation.

We considered two basic possibilities. Mutants that ensure that individuals "stay together" are a critical part of the evolution of eusociality (and, when it comes to the cells that make up creatures like you and me, of multicellularity too). If a gene makes offspring stay with their mother and help her—it could, for example, be a disruptive mutation in a gene that would normally make the offspring leave to set up their own nests—then we are dodging the Prisoner's Dilemma. In this case, the workers are not independent agents. Their properties are determined by the genes that are present in the queen (both in her own genome and in that of the sperm she has stored). The workers can be viewed as "robots" built by the queen. They are part of the queen's strategy for reproduction. This is not a cooperative dilemma, or even an evolutionary game.

There is a second possibility, wherein a gene makes individuals "come together." In this encounter between players, we typically have a cooperative dilemma. For example, several fertilized queens cooperate to establish a new colony, which does indeed occur in the case of some ant species. These two mechanisms should be borne in mind when developing a theory for the evolution of eusociality.

SUPER NATURALIST

Wilson, the lord of the ants, grew up in rural Alabama, the only child of a government accountant. In 1936, when he was six years old, his parents divorced. He can vividly remember his encounters with wildlife at that early age. He studied a jellyfish one afternoon in the crystal waters of Perdido Bay as it floated absolutely still in the water. He had never dreamed of any such thing before, and so the suspended jellyfish, a sea nettle, came to symbolize "all the mystery and tensed malignity of the sea." He was eager to know what else lurked in that deep, mysterious chamber of blue, speared with light.

When he was seven, he had an accident that, as he put it, "determined what kind of naturalist I would eventually become." This mishap occurred at Paradise Beach, near Pensacola, Florida, fishing on a dock for pinfish. He jerked one out the water and one of the needlelike spines on its dorsal fin penetrated the pupil of his right eye. Eventually,

the lens of the eye had to be removed in what was, at that time, a terrifying ordeal. Fortunately, he had close-up vision in his left eye, perfect
for seeing the hairs on the body of an insect. He was now "committed
to minute, crawling, and flying insects, not by touch of idiosyncratic
genius, not by foresight, but by a fortuitous constriction of physiological ability." The little pinfish had turned him into an entomologist.

Wilson also likes to joke that every kid has a bug period. "I just never
grew out of mine." As a ten-year-old, while exploring the National Zoo
and nearby Rock Creek Park in Washington, he became enthralled by
the "magic world" of insects. At the age of thirteen he made his first
publishable discovery, a species of fire ant in Mobile, Alabama, that
subsequently spread throughout the South. He would go on to study
for his bachelor's degree in biology at the University of Alabama and
then move on to his master's.

By then, his mastery of all things ant was apparent. His writings on
the little-known dacetine ants prompted an entomologist to urge Wilson to transfer to Harvard, home to the world's largest ant collection.
There he studied the social behavior of ants, which he went on to show
was influenced by chemical signals. Wilson can recall how, one day in
1959, he removed the Dufour's gland from a fire ant, crushed it, and
smeared it across a piece of glass. The fallen ant's fellow workers rushed
forth, following the smear to its end, where they milled around. The
gland was clearly the source of a pheromone, one of a group of chemical substances secreted by ants to signal food, danger—even death.

In the next decade Wilson came across Bill Hamilton's work on kin
selection, which stirred his interest in what mathematics had to say
about the world of the ant. Wilson was at that time highly receptive to
the idea of a kind of Newtonian law of biology and, through his avid
support, helped to establish kin selection as a dominant theory (which
is why his current opposition to the idea is all the more striking). He
had worked hard to put more mathematics into the discipline at the
start of the 1960s with other thrusting young population biologists,
such as Richard Lewontin, feeling that mainstream biology was lag-
ging behind the amazing advances being made at that time in molecular biology.

Wilson recalls how, in the spring of 1965, he first read Hamilton's paper on a train ride from Boston to Miami. As he set out from Boston, he was skeptical. But he gradually warmed to the idea during the following eighteen hours he spent in his roomette. By the time his train finally rolled into Miami, he was a convert to Hamilton's dazzling "haplodiploid hypothesis" that initially gave kin selection its magnetic power. "It was brilliant," says Wilson. "I still think it is." His backing would catapult kin selection into mainstream discussions.

Here's the essence of the hypothesis. Females develop from fertilized eggs, while males develop from unfertilized ones. As a result, females are diploid (they have two copies of the entire genetic code, or genome, just like people). Males are haploid; they have only one copy of the genome. This method of determining sex, called haplodiploidy, ensures that sisters are more closely related to each other than to their own offspring. And this means that the best chance they can give their own genes of surviving is to look after each other rather than lay eggs of their own.

The haplodiploid hypothesis works as follows. In a haplodiploid species sisters are 75 percent related, but mother and daughter are only 50 percent related. Hamilton's rule predicted that hapolodiploid species would prefer to help raise their sisters rather than produce their own daughters. This is what provides the stability at the heart of the ant colony. Other insects also use haplodiploidy as the sex-determining mechanism, notably bees and wasps.

Wilson was "enchanted by [the idea's] originality and seeming explanatory power." From the occurrence of hapodiploidy alone he could draw a number of conclusions: societies of altruistic sisters might be expected to evolve more frequently among the ants, bees, and wasps than in others with conventional diploid sex determination, where both sexes have two sets of chromosomes. This seemed to be the case with Hymenoptera, an order of insects comprising the sawflies, wasps, bees, and ants, though it did not seem to fit the termites.

Unlike mathematicians, biologists do not dismiss an idea if there is a single exception. Indeed, they often resort to the baffling motto that declares that "one exception proves the rule." Perhaps the insights that the rule can provide are so beautiful and intoxicating, it seems a shame to let an ugly and inconvenient fact get in the way. With kin selection one could, for example, predict that the worker castes of these species should be female and that males should be drones, whose sole function is to mate with the queen. The haplodiploid hypothesis felt like a magical key that could unlock new mysteries as effectively as the recently found structure of DNA.

In the fall of 1965, Wilson sailed on the *Queen Mary* to England to give an invited lecture on the social behavior of insects to the Royal Entomological Society in London. He remembers wandering around the great city with Hamilton, then a mere graduate student, telling him not to be discouraged by the initial indifference that had greeted his ideas about inclusive fitness. The two ended up promoting Hamilton's work at the meeting of the society. Few of the panjandrums present were familiar with the Hamilton paper, and they were skeptical as a consequence. But Wilson was prepared. He had already thought about many of the objections. With Hamilton, "we carried the day."

FROM ANTS TO SOCIOBIOLOGY

By the end of the 1960s, Wilson felt that the time had come to draw together the many threads of experimental and theoretical work on social insects. He wanted to create a synthesis of everything known about them that would spin "crystal-clear summaries of their classification, anatomy, life cycles, behavior and social organization." Driven by what he called the "amphetamine of ambition," he resolved to write a book on a discipline that he decided to dub sociobiology. Wilson's research for his 1971 book *The Insect Societies* led him to the belief that behavior might result from genetic evolution, rather than from learning or cultural forces.

Wilson hoped that the fecund new gene-based ideas could be extended to provide the basis for understanding the evolution of social behavior, whether in social insects or social vertebrates, from murders of crows to flocks of sheep. At the back of his mind, he also thought that the concepts were powerful enough to apply to people (as he would put it, "Let us now consider man in the free spirit of natural history, as though we were zoologists from another planet"). By unpacking this explosive idea in his book *Sociobiology: The New Synthesis* in 1975, Wilson ensured that Hamilton's kin selection theory penetrated deeply into the public consciousness.

In his book, Wilson pioneered and popularized attempts to build on this theory in order to explain the evolutionary mechanics behind behaviors such as aggression, altruism, promiscuity, even division of labor between the sexes. Although voted the most important book on animal behavior of all time by officers and fellows of the international Animal Behavior Society, *Sociobiology* became the object of bitter attacks by social scientists and other scholars, even from his former collaborator Richard Lewontin.

In a letter to *The New York Review of Books,* Lewontin and the Harvard evolutionary biologist Stephen Jay Gould were among a dozen professors, doctors, and students who condemned *Sociobiology* as providing "a genetic justification of the status quo and of existing privileges for certain groups according to class, race, or sex." The left, sensitized by earlier false arguments about racial science, loathed the idea that human social behavior, indeed human nature itself, has a biological foundation. They feared that this kind of thinking was politically dangerous, the kind of idea that led to the establishment of gas chambers in Nazi Germany.

Wilson was denounced as a racist, a sexist, and a fascist (he is a Democrat, and a good-natured one at that). At one meeting of the American Association for the Advancement of Science, he was dowsed with a pitcher of ice water by demonstrators as they chanted, "You're all wet!" At another meeting, this time of the American Anthropological Association, delegates considered a motion to censure sociobiology. Looking back on the furor, Wilson remarks that the book "was a hand grenade with the pin pulled out."

Wilson resolved to answer his critics. In his next book, *On Human Nature,* he argued that most domains of human behavior—from child care to sexual bonding—are the result of deep biological predispositions consistent with genetic evolution. The danger of oppression lay not in sociobiological theory, but in uninformed views of man's evolution— in particular, the kind of genetic pseudoscience that led to restrictive immigration laws in the United States and to the eugenic policies of Nazi Germany. The book won a Pulitzer Prize in 1979. But Wilson would return to his beloved insects, coauthoring a definitive 732-page book, *The Ants*. A second Pulitzer followed in 1991.

SPRING-LOADED SOCIETY

Hamilton's dazzling haplodiploid hypothesis held true throughout the 1960s and 1970s, when it gave kin selection the same glossy sheen as a fundamental law of physics. But by the 1990s, the hypothesis began to dull and then to fail. At first, the termites were the lone, troublesome exceptions. Thanks to the efforts of one of Wilson's students, Barbara Thorne at the University of Maryland, they were found to provide a good fit for group selection ideas instead. Then Wilson encountered the work of James Hunt at North Carolina State University and Raghavendra Gadagkar of the Indian Institute of Science, Bangalore. Both are wasp experts and both concluded that kin selection did not fit their observations.

At the same time, examples have emerged of eusocial creatures that are diploid rather than haplodiploid in sex determination. These include the platypoid ambrosia beetle, synalpheid sponge-dwelling shrimp, and bathyergid mole rats. Overall, half of all eusociality seems to have arisen in such diploid lineages. There are also societies that seem to have all the right ingredients to be eusocial but are not. For example, among the 70,000 or so known parasitoid and other apocritan Hymenoptera, one of the largest orders of insects, all of whom are haplodiploid, no eusocial species has been found. Nor has a single example come to light from among the 4,000 known hymenopteran sawflies and horntails, even though their larvae often form dense cooperative aggregations. Unsurprisingly, the haplodiploid hypothesis has now been abandoned by many who study social insects. Wilson often asks kin selectionists why they still cling to this idea: "They like to say, 'Well, why bring that up?'"

Wilson's current view about the origins of eusociality is profoundly different from the assessment in his seminal book *Sociobiology: The New Synthesis*. According to that widely accepted earlier account, selection acting on individuals that are related (kin selection), rather than on whole colonies, explains the origins of eusociality. Today, Wilson does not emphasize kin selection. Instead he focuses on ecological factors and on genes that predispose insects to colonial life.

By Wilson's reckoning, the genesis of the ant colony begins with the nest. He points out that all of the branches or groups (technically called clades) known to have primitively eusocial species—aculeate wasps, halictine and xylocopine bees, sponge-nesting shrimp, termopsid termites, colonial aphids and thrips, ambrosia beetles, and naked mole rats—rely on colonies that build and occupy defensible nests. In a few cases, unrelated individuals join forces to create the little fortresses.

In most cases of animal eusociality in the Hymenoptera, the colony is begun by a single inseminated queen. Before the emergence of eusocial insects, a solitary insect species would reproduce by what Wilson calls progressive provisioning. Mated females build a nest, lay eggs, and feed the larvae. When the larvae hatch, the offspring leave the nest. Wilson and I believe that crossing the threshold to eusociality requires only that a female and her adult offspring do not depart to start new, individual nests but instead stay put. The high relatedness in the resulting colony is better explained as the *consequence* than the *cause* of eusociality. Once eusociality has evolved, colonies consist of related individuals, because daughters stay with their mother to raise further offspring.

Of course, for these extraordinary societies to evolve, there has to be an advantage to being social. The mathematical analysis shows that the fundamental question is how are the key demographic parameters of the eusocial queen (her fecundity and her risk of death) affected by the presence of workers? In the presence of workers, the eusocial queen has two fitness advantages over solitary mothers: she has increased fecundity—birth rate—and reduced mortality. While her workers forage and feed the larvae, she can stay at home, which reduces her risk of predation, increases her rate of egg laying, and enables her (together with some workers) to defend the nest.

We found that it is easier to maintain eusociality than to evolve it. For a wide range of circumstances it was unlikely that a eusocial mutation would take hold in a solitary society. Equally, however, once eusociality had evolved it could no longer be displaced by a solitary lifestyle. This explains in part why, even though eusociality is ecologically dominant, the condition has evolved rarely in the history of life.

What follows once a proto-eusocial nest has been established? The offspring of the queen would possess preexisting features—a "behavioral ground plan"—that could spring-load eusocial life. These traits include progressive provisioning of larvae, by which many can be reared at the same time, and flexibility in behavior that allows division of labor. For example, the females of solitary wasp species lose some of their young to predators when they forage for food, but if a second female were available to serve as a guard, these losses could be cut.

In the earliest stage of eusociality, the offspring that stayed with the nest would be expected to assume the worker role, following a preexisting genetically encoded behavioral rule. There is already good evidence of how such rules could spur cooperation—for instance to help care for offspring. Wilson likes to cite a beautiful Japanese experiment on solitary bees that were forced to build a nest on the same spot. Bees are programmed to make a nest in stages, so that when one had finished one construction job the second built on those efforts. At the end of their enforced partnership, one bee descends to the bottom of the nest—whether this is luck or dominance is not known—to become a queen and lay eggs. The second bee, seeing that the chore of producing offspring has been completed, takes on the role of foraging. Coerced partners have been forced to divide labor in guarding, tunneling, and foraging too.

As societies grow larger and more complex, however, competition between colonies grows fiercer, and as a consequence, group selection begins to act, spurring the emergence of a worker caste selected from genetic mutations within a group. This origin of an anatomically distinct worker caste appears to mark what Wilson calls the "point of no return" in evolution, at which eusocial life becomes irreversible. It is at this stage that insect societies made the transition to superorganisms. Eusociality, like multicellularity, is an important invention in evolution, one that shows the incredible power of cooperation.

GENESIS OF SUPERORGANISMS

Superorganisms arise from the initial formation of groups, along with a minimum and necessary combination of preadaptive traits, such as the creation and defense of a nest. Mutations guaranteed the persistence of the group, most likely by preventing ants from leaving the nest. As a result of what Wilson calls "spring-loaded preadaptations," a primitive form of eusociality was established. This was honed and elaborated as emergent traits caused by the interaction of group members were shaped through natural selection by environmental forces. Finally group-level selection drives changes in the colony that result in sophisticated features such as fungi gardens and aphid herds.

Given the discussions about inclusive fitness in chapter 5, I should underline why this model is different. With our new mathematical model we see very clearly why it is not simply "relatedness" that drives the evolution of eusociality. In many solitary species there may have been mutations that make daughters stay with the nest. But whether or not one such species makes the step toward eusociality cannot be explained by relatedness alone because in all of those attempts the daughters are equally related to their mothers. Instead, it is down to parameters of life history, such as an increased rate of laying eggs and an increased longevity of the queen in the presence of workers, that really determine whether or not natural selection favors eusociality.

Finally, it is fascinating to contrast two-legged and six-legged society. Both owe their success to cooperation and division of labor. Both rely on multilevel selection, where there has been competition between groups. But, of course, ants are ruled by instinct alone while, thanks to language, we also have swiftly evolving cultures. Before we feel too smug, we should remember that after only 200,000 years, we humans are in danger of overwhelming our planet while the ants have lived in harmony with it for 100 million years.

Wilson likes to point out that both our civilization and that of the leaf-cutter owe their existence to agriculture. What is remarkable is that while our relationship with plants catapulted our species out of its hunter-gatherer lifestyle around 10,000 years ago, some social insects had achieved this transition 60 million years earlier. There may be parallels between the scenarios of animal and human eusocial evolution, and they are, we believe, well worth examining to shed light on how we made the step from wandering tribes of hunter-gatherers to hamlets, towns, and cities. Queen leaf-cutter, who rules over the greatest super-organism, still has much to teach us.

From Cooperators
to SuperCooperators

CHAPTER 9

The Gift of the Gab

*What a piece of work is a man, how noble in reason, how infinite in
faculties, in form and moving how express and admirable, in action how
like an angel, in apprehension how like a god! The beauty of the world,
the paragon of animals—and yet, to me, what is this quintessence of dust?*
—William Shakespeare, *Hamlet*

High thoughts must have high language.
—Aristophanes

Gossip. Banter. Chat. Let's talk. Let's organize a colloquium. Even
better, let's have a party! Language allows people to work together to
exchange their ideas, their thoughts, and their dreams. In this way,
language is intimately linked with cooperation. For the mechanism of
indirect reciprocity to work efficiently it needs gossip, from names to
deeds and times and places, too. Indirect reciprocity is the midwife of
language and of our big, powerful brain.

The birth of language is perhaps the most amazing event to occur
in the last 600 million years, one that is of equal significance in the
unfolding drama of evolution to the appearance of the very first life.
That is because language provided a vast new stage upon which Darwin's struggle for existence could play out, a novel mode of evolution

and a remarkable spur for cooperation, even among people who are separated in time and in space.

Thanks to the gift of the gab, humans now occupy a unique place in the 4-billion-year story of life on Earth. Before they arrived, the most significant way that information was transmitted between living things was in the form of the chemicals of heredity, DNA or RNA. Then along came language, an ever-growing repertoire of signals that had evolved from ancient regions of the primate brain that were once only used to decipher sounds and control facial muscles. In this way, language propelled human evolution out of a purely genetic realm, where it still operates, into the realm of culture.

Language offers a way to take the thoughts of one person, encode them, and insert them into the minds of others. We can achieve that remarkable feat of thought transference without the pain, drilling, and bloodshed of trepanning. If someone has a great idea, it can spread instantly and does not have to wait for fission, sex, or infection to propagate its influence down the generations. With the emergence of *Homo sapiens,* units of spoken and mental information now began their own strategies for selfish proliferation and cooperation. It was William Burroughs, the opiate-fueled writer, who memorably suggested that language itself might be a virus, manipulating us for its own purposes. This virus has led to the dramatic increase in the speed of change on the planet, for better or for worse.

Language became a spur for evolution. Those who had brains that were most receptive to these new ideas, and could make best use of them, were more likely to thrive. The fittest in this cultural sense could be the one who is most imitated, leaves the most disciples, or whose philosophy or technology has been adopted the most widely. Language, in effect, helped bootstrap the development of our powerful and flexible brains. We like to think that we created language, but this is back to front. Language created us. Locating the origins of language could help to shed light on the origins of humanity.

But tracing the trickle of grunts and other utterances that lies at the source of this mighty river of sounds is fraught with difficulty. Words leave no petrified remains. If only the damp rock walls of a mossy cave could carry the faint imprint of conversations that were held there long ago. If only there was a way to extract ancient sounds from the sand, pebbles, and stones that litter a prehistoric settlement. What an incredible story they would tell! A tale of mammoth hunts, of the struggles against rival tribes with flint weapons, the demise of our cousins the Neanderthals, the rise of agriculture, cities, and civilization and so much more. But there's no record. Not a trace. Not a single acoustic fossil to reveal any syntactic precedents, nor how language came into being. We cannot hear the voices of the dead.

I believe that the solution to the mystery of language's beginning lies not only in linguistics but in understanding our ancient origins. Language had to evolve with cooperation, since individuals are not going to bother to learn new ways to communicate with each other unless they are already working together to some extent. Associations between noises that we can make and the meanings we want to communicate can only knit and form when information transfer is beneficial for both speaker and listener. In this way, language and cooperation coevolved. Ugg looks at Igg and points toward the waterhole where an antelope is drinking. Igg nods, picks up his newly sharpened flints, and joins him in the hunt. This little act of cooperation can help them both to thrive with fresh steak that night. Equally, language could make cooperation work in new ways. Now people could help each other without even meeting directly. Ugg could tell Igg that he saw several antelopes at the waterhole yesterday evening. The herd might return tonight and Igg should also ask his friend Agg to come along for the hunt.

Using this as my starting point, I tried to work out why language is designed the way it is: divided up into units, called words, that are arranged by rules we call grammar. I wanted to use mathematics to connect evolutionary ideas with ideas about learning and about language theory, a highly abstract zone of human thought where linguistics and computer science embrace. To put the momentous event that gave us songs, prose, drama, and so much more on mathematical foundations, I needed a model of the way language evolves among cooperating individuals.

I would be the first to admit that the great puzzle of language evolution is by no means easily solved. This subject is so rich and deep that, for every problem we cracked, dozens more have appeared. New lines of inquiry take root, sprout, and then grow. This new challenge also came at a time of major upheaval in my life. Nonetheless, it was a most amazing journey.

INSPIRATIONAL DUCK

In 1997 I was professor of mathematical biology in the Zoology Department at Oxford and senior research fellow at Keble College. Though some find Keble's polychromatic brickwork an acquired taste, I adore the Gothic style of William Butterfield, who claimed that he "had a mission to give dignity to brick." My work on the evolution of cooperation was sent spinning off in a new direction when a colorful individual came around for dinner. David Krakauer had been born in Hawaii, raised in Portugal, and now found himself in Britain, working in the Zoology Department. He was passionate, energetic, and driven by an insatiable curiosity.

David came to visit almost every night. And almost every other night I wanted him to roast a duck. This was a very personal example of direct reciprocity. David's duck recipes were flawless, beyond comparison. And I wanted to match his efforts by being an ever grateful and appreciative diner. He always insisted that we buy the best ingredients. The covered market in Oxford was a good place to find fine poultry, with lots of choice—French Barbary duck, Aylesbury duck, or Gressingham duck? And only the very best cherries would be good enough to make a sour compote to cut through the crispy skin and fattiness of the meat. We slavishly followed his mantra: "With duck, you cannot take any shortcuts."

One evening, as a duck was once again roasting in the oven, David let slip an idea, a way to probe the origins of language, a neat approach to capture its quintessence so that language could be examined under the microscope of mathematics. It all boiled down to representing communication with a matrix. This way of showing the relationship between various data is familiar in everyday life. Think of charts that show fares and destinations, railway timetables showing the times that an express train arrives in each place, tables that show investment periods and interest rates, the cooking times of a given weight of duck, and so on. In the matrix that David devised, on one side he had signals and on the other he listed objects. This simple idea would germinate a seed of curiosity.

I was fascinated and wanted to formulate a language game around this idea, one that could shed light on the origins of language. I had the same visceral feeling I'd had when Karl mentioned indirect reciprocity in the Wienerwald. I felt something new and great would come out of this idea. In fact I felt it was inevitable. But before I could make a start, my career path in academia would undergo an extraordinary change.

OXFORD DUSK

Leaving Oxford was the last thing on my mind. I adored the city of dreaming spires, its Eights races on the Cherwell River, gowns and other traditions, the local countryside, and the playful academic approach. There were many sunny afternoons when I played tennis on the grass courts in the university park with the geneticist Richard Moxon; Jim Watson of double helix fame; the bishop of Oxford, Richard Harries; and many others. The list, of course, includes Bob May, who managed to beat me in our very first encounter. By now Bob was the prime minister's chief scientific adviser, the most influential job in British science. Despite the heavy demands of his position, Bob's curiosity remained unquenchable. Every day as soon as he got back from Whitehall he would flop down in my office and ask, "What's up? What do you have?"

One day, however, Karl phoned to say that my work had created a little buzz at Princeton's Institute for Advanced Study. I was intrigued, though I did not read too much into this piece of intelligence. Years before I had read *Who Got Einstein's Office?*, in which Ed Regis gave an effervescent account of life there, which seemed well beyond the reach of a mere mortal. This was an extraordinary institution with "ethereal pretensions" that had been populated over the years by the high priests of science and mathematics, the likes of Albert Einstein, Kurt Gödel, and John von Neumann.

Yet a few months later I was made a remarkable offer to join the institute. Phillip Griffiths, who was the director at that time, invited me to head the institute's first program in theoretical biology. The initiative was to be supported by the New York philanthropist Leon Levy. This was an incredible opportunity.

The reaction from my peers was mixed. Bob studied the details of the institute's offer and concluded that it was pretty much the most that anyone could expect. He was very happy for me and said that I should leave. He would have done the same himself. John Maynard Smith told me not to go. He had felt intellectually isolated when he had once visited this monastery of the mind. "There was nobody to talk to," John said. Sir Richard Southwood, who was the head of the Zoology Department when I arrived and later vice chancellor of the University of Oxford, also told me not to leave. Everything was going so well for me at Oxford and, he reasoned, who knew what would happen if I moved to Princeton.

I was still sitting on the fence, albeit with both legs dangling on the Princeton side. There was one last hurdle. I had to meet the new department head at Oxford, Roy Anderson, who was very charming and very persuasive. I feared that Roy might convince me to stay. But when I entered his office, he shook my hand and wished me the best. That was it. Both a sad and a great moment. Roy had always given me a lot of support, and it later emerged that Bob had asked him not to prevail on me to reject the Princeton offer.

The last time that I saw Bob before I left for good was steeped in nostalgia and emotion. He handed me a book on applied mathematics, *A Course of Modern Analysis* (1902), by Edmund Whittaker and George Neville Watson. He had inherited the classic tome from his adviser, Robert Schafroth, who had died in a plane crash. "This," he told me, "contains many of the tools that I use for work." Bob had inscribed the book "From Robert Schafroth to Robert May." Next to "Schafroth," Bob had written "First to observe that charged bosons superconduct."

Then, underneath, he added, "From Robert May to Martin Nowak." We were both moved to tears.

LANGUAGE, LIFE, AND PRINCETON

My relocation from Oxford to Princeton marked the start of a thrilling new venture. My research on the Second Big Bang of biology could now begin. Some people had proposed that language had emerged as the consequence of a bigger brain. I didn't buy this idea. Language is something extremely specific. Just making the brain a bit bigger doesn't give you the ability to generate language. In fact, I believe that the relationship is the other way around: as evolution selected individuals who could communicate with improved language, it selected for bigger brains at the same time.

I found myself working on the problem in a little refurbished house at the edge of the forest around the institute. Working with me were Ramy Arnaout, David Krakauer, Alun Lloyd, Karen Page, Joshua Plotkin, Lindi Wahl, and Dominik Wodarz, all friends from Oxford. But life was so different from Oxford, where bikes were the preferred form of transport. Alun consistently rejected the convenience of the automobile and walked everywhere, whether five miles to the supermarket or back home. This caused surprise among local people, even consternation. He was bemused to find himself being stopped by the local police who suspected him of "loitering with intent."

Perhaps to underline the way in which biologists were regarded as outsiders at the institute, our wooden house was on the perimeter of the campus. We worked opposite the nursery school, a one-story brick building that had once housed John von Neumann's pioneering Electronic Computer Project. As soon as von Neumann had died, the institute (always suspicious of anything the least bit practical) had donated the computer to a museum, the Smithsonian Institution.

My sons went to that nursery school and learned language from their teachers, along with their American and international peers. While they were conducting their experiments on reading and writing, I was sitting a few meters away, engulfed in the theory of language acquisition.

History seeped from the walls. My office was once occupied by Julian Bigelow, the electrical engineer who had built von Neumann's machine. Gödel had lived in a small house nearby and his extraordinary and paranoid presence could still be felt, both in my work and in the endless stream of oddball anecdotes from those who had crossed paths with him. Freeman Dyson, the British theoretical physicist and visionary, had told me how a paranoid Gödel had once called him from his office explaining that he had just received a big, light parcel. Gödel was fearful the parcel contained poison gas and had implored Dyson, "Would you be so kind as to open it for me?" Yes, came the reply to this highly illogical proposition. Gödel, though grateful, had insisted that Dyson open it alone and in his own office. Inside was a beautiful paper model of a mathematical surface.

Much of the work at the institute was abstract and idealized, focusing on the Platonic forms of mathematics that are unchanging, perfect, and eternal. I was also a Platonist. But my work was a little heretical, from the perspective of my fellow monks. I was using mathematics as a scalpel to dissect something that was perceived to be changing, slippery, yet very tangible. I wanted to explore the Platonic truth and beauty of language and its evolution.

The evidence backing evolution is overwhelming. Yet there has always been unease about extending Darwin's revolutionary thesis from evolving genes to evolving language. The Société de Linguistique de Paris officially banned all work on language evolution in 1866, just after Darwin's ideas were published. Mathematicians love a challenge and, for me, this amusing historical fact made the study of language evolution even more compelling.

Even today, however, many linguists, biologists, and philosophers have difficulty imagining how language could have arisen through evolutionary forces. Sure, language emerged because it helped our ancestors to share information that was somehow crucial to survival. But, as Harvard University's Steven Pinker has pointed out, such explanations seem glib and circular. "Bad [explanations] try to explain one bit of our psychology (say, humor) by appealing to some other, equally mysterious bit (laughing makes you feel better)." Of course language aided survival— "Mind that spear, dear!"—but if it were that simple, gorillas would be having highbrow discussions about the origins of language too.

While I thought that my work on cooperation had yielded up some tools to understand language evolution, even the great Noam Chomsky had shown some skepticism that natural selection could ever shed light on the origins of language. Chomsky is seen by many as the Einstein of linguistics who, since the 1950s, has pioneered the study of language's complexities with surgical precision. Half a century ago, he established a research agenda that would exert a grip on linguistics, psychology, and computer science for decades to come.

The key to understanding grammar, the most fascinating ingredient of language, had come from Chomsky. One can think of it as a way to take linguistic forms, words, and arrange them so they have a specific meaning. Grammar, if you like, enables words to cooperate to form new meanings. Grammar is what gives rise to the unlimited expressibility of language.

Chomsky argued that language is, at its heart, a matter of encoding and decoding strings of arbitrary symbols. By formulating language in this basic way, he could address a central problem: How does the human mind, with its finite means, make unbounded use of symbols in very specific and organized ways to communicate? He found that an elegant mathematical underpinning can generate a code to speak and understand infinite meanings. Chomsky argues that all languages have a common structural basis, a set of rules that he called "universal grammar," a plan common to each and every grammar that we know of.

THE BIG BANG OF LANGUAGE

We can't say precisely when the first language emerged. It could have been any time between 7 million years ago, when we shared our last common ancestor with chimps, and the emergence of anatomically modern humans 150,000 years ago. But in what circumstances? What kind of communication did our ancestors use in their hunting and gathering? David Krakauer and I began by examining the "primordial soup" of language that is present throughout the animal kingdom.

We can see the ingredients all around us, in the chemical signals that flicker on and off between cells, the waggle dance of bees, territorial calls of various creatures, and the remarkable repertoire of birdsong. There are the complex patterns of moans, cries, and chirps in the songs of whales. There are ring-tailed lemurs that waft their tails around to speak a complex "language" of multiple scents to communicate aggression, receptiveness to mating, and much more. Let's not forget those extraordinary intellects of the marine world, the jet-powered cephalopods, whose rapid changes in skin color and pattern are the basis of courtship rituals. And then there are our closer relatives, such as the chimpanzees.

Any way you look at it, the languages used in the rest of the animal kingdom fall short of our abilities. Most animals use nonsyntactic communication, so that a single grunt—a "word"—might be used to express a scenario, such as "watch out—there's a lion prowling nearby." Putty-nosed monkeys (*Cercopithecus nictitans*) in West Africa use two main sounds—*pyows* and *hacks*—to warn each other about predators. Of course, a language does not even have to be built on a foundation of sounds. Scouts from beehives perform dances inside the nest to tell their peers where nectar can be found. The coordinates of distant locations are encoded in the waggle phase of this buzzing ballet, with the direction and distance to the food source indicated by the orientation and duration of the dance.

Leaving aside the tricky issue of grammar, which I will return to later, these kinds of considerations allowed me to comprehend that the size of the repertoire of signals we use—the lexicon—is a critical issue. When it comes to our repertoire, a six-year-old has a lexicon of about 13,000 words. The rate of word learning in humans comes to about one word every ninety waking minutes from age one to age seventeen. This leaves a seventeen-year-old native English speaker with about 50,000 words stored in her mental lexicon, which is the typical value for a grown person.

Absorbing this vocabulary is a colossal task, similar to learning 50,000 telephone numbers with all kinds of associations. We don't realize what wonderful memory machines we are. But memory is not the whole story. The human vocal tract can make a variety of different sounds. Based on the six thousand languages that we know of, our tract can make around 1,000 linguistic sounds. We call these units phonemes and the total inventory in languages varies from as few as 11 in Rotokas, which is spoken in Bougainville, an island to the east of New Guinea, to as many as 112 in !Xóõ, a "click language" of Africa, notably Botswana and Namibia. These may range from tones to familiar sounds to clicks, which can only be written by using the sort of characters you usually press by accident on a computer.

Another idea naturally follows from this physiological fact. In speaking this vast lexicon, we make mistakes. Producing the sounds—the phonemes—in an ordinary conversation is a remarkable anatomical feat. The motions of various parts of our vocal tract are coordinated within millimeters and timed within hundredths of a second. The next time you open your mouth to speak, just remember how breathtakingly awesome you are. But errors are inevitable; and, in essence, the more concepts that have to be communicated, the more sounds that are needed to do this, the closer the sounds will be to each other, and the greater the risk that this repertoire will become confused.

Thus there is a limit to the number of phonemes that we can handle with our vocal apparatus. If we used the simplest kind of language, where one phoneme is linked to one action, object, person, or whatever, we would quickly run into difficulties as we developed more associations. Think about the potential for confusion and disappointment if only slightly different sounds are used to distinguish the discovery of a cache of ripe bananas from a fly-eaten pile of rotten black ones. There is a wonderful Gary Larson cartoon that shows two apes dancing a tango with each other. The caption reads: "I'm afraid you misunderstood . . . I said I'd like a mango."

To capture the physiological reality of speaking a language, I proposed the idea of a "linguistic error limit": the number of distinguishable sounds in a protolanguage, and therefore the number of objects that can be accurately described by this language, is limited. Adding new sounds increases the number of objects that can be described, but this flexibility comes at the cost of an increased probability of making mistakes; the overall ability to transfer information does not improve.

The error limit that we calculated suggests a reason why the overall number of phonemes we use in any one language falls far short of the possible range of around 1,000 sounds that can be made by the vocal tract. If there are too many, there is too much potential for confusion. Better to concentrate on a few phonemes and learn how to distinguish them effectively. Equally, once we have learned one set, it can be hard to change to another. That is why foreign accents, like my Viennese take on English, are typically caused by the use of the wrong phonemes.

THE LANGUAGE GAME

Armed with this understanding, we can now begin to map out the origins of human language. At the dawn of the spoken word, our ancestors had a few utterances, from howls to snorts and grunts. Then specific sounds became associated with specific objects. Ugg points, makes sounds, and his audience, Uggette, thinks, "Aha, that is what Ugg is trying to tell me." Crucially, however, cooperation is needed for this to occur.

The speaker and the hearer must have some common goal, some shared venture, from sex to steak. Otherwise why would I get a payoff for successful communication? Uggette and Ugg have to turn their words into deeds. Associations between a particular sound and a specific meaning will only gel when information transfer is beneficial for both speaker and listener, for both Ugg and Uggette. Then, as a society attempts to describe more objects, more sounds are needed.

We began to model this with the matrix that David first proposed over that duck dinner in north Oxford. We took an idealized simple signaling system where a "referent"—say the appearance of a cheetah lurking in the bushes—produced a signal, a warning cry of "Get out!"

The hope is that, for a second person, that warning cry would produce the same referent, in other words, trigger the alarming thought "A cheetah is coming!" I wanted to know what would happen in a population of animals playing a game where there are initially random associations between referents and signal, as each individual adopts his or her individual cry of alarm for that approaching cheetah. Over evolutionary time, would consistent associations occur so that the same word was used to describe the cheetah?

We set up a simple game, a coordination game, where points were awarded for successful communication. As ever, we used a computer to compress the passing of the eons. In our model, we looked at groups in which random signals were used to talk to everyone else in a group and points awarded if individuals successfully cooperated. Although this sounds too simplistic, there are more complex versions of this game that can deal with deception. There are endless examples of this in the wild, such as the monkey who was observed making a warning call at the very moment that his fellow primate had just found a ripe banana, prompting the latter to drop his prize so it could be stolen. But the essential gist of what is going on still comes through even in the simplest form of the game, where the individuals with the best payoff end up with the most offspring.

What is important for the evolution of language is the "ecology" in which an animal operates. Which predators are most dangerous? What food is available? How is it obtained? Does it require cooperative efforts to go hunting (or gathering)? How many creatures can band together in groups in your species? What kinds of interactions take place between individuals: grooming, courtship, fighting for dominance, and so on and so forth?

When cooperation emerges, then it is not enough to be the biggest, strongest male. Your smaller peers can gang up on you and, in this way, politics is born: "Now, hang on guys—let's not fight. Let's be reasonable and talk about this." Equally, one can cooperate with one's peers to hunt bigger, more dangerous game. Or one could see how rich localized sources of food on the forest floor could encourage ants to build fortresses nearby. It is the "ecology" that drives the need for solutions to myriad problems and spurs on evolution. Language is one such solution.

We found that maximum fitness—the Darwinian shorthand for success—is achieved by using a small number of signals to describe the most valuable objects. Adding more signals reduces fitness. Let's say a monkey has three different words for the three different predators that eat it. Then it is dubious that having another word for a relatively harmless creature such as a zebra in its lexicon is worth the effort and worth the risk of possible confusion. We came to realize that evolution has to overcome this limit, where the addition of signals has the paradoxical effect of diminishing understanding.

In the second step of language evolution, driven by their richer social ecology, humans left animals behind. Our ancestors overcame the error limit not by forming more sounds but by combining a few easily distinguishable sounds into words. In this way, meaningless vowels and consonants such as b, a, and t can be assembled into a meaningful word "bat." As a bonus, they can form "tab."

I showed mathematically how such word formation enables a language to convey a colossal number of objects. We can shuffle the phonemes on two levels, first combining strings of phonemes into words and then words into sentences, which is called the duality of patterning. The creation of sentences requires the last step in language development, the incorporation of rules in the form of what we call grammar. Now a limited diversity of words can be combined in essentially an unlimited way. Thus grammar enables us, in the words of Wilhelm von Humboldt, to "make infinite use of finite means." This feat of creativity through recombination sets us apart from the rest of nature.

To illustrate the effect of grammar, take the example of a foreign recipe and foreign gossip. In terms of linguistic complexity, recipes are simple. Try this out yourself. Use a foreign language recipe book and, so long as you know what the words mean, it is relatively easy to figure out how to cook a feast. But if you overhear a conversation and hear the words "ball," "Roger," and "stranger," it could either mean that Roger is asking a stranger to a ball or is asked by a stranger to come to a ball, or they are about to play a ball game. Here the word order is all-important for sense. Indeed, the meaning of the word "ball" is also context dependent. This sort of conversation uses the grammatical structure of language to the full.

In the same way, the words "man" and "bites" and "dog" can be combined in two quite different meanings. One, according to traditional Fleet Street lore, is a story that should grace the pages of a self-respecting newspaper. It is indeed news when a man sinks his canines into a dog. The alternative, a wounded man with bloody teeth marks on his leg, is just a humdrum consequence of our ancestors having domesticated grey wolves fifteen thousand years ago. In this way a rare-but-remarkable event, one that might merit a few paragraphs in a website or a newspaper, can be easily described by combining words in ways that need not be learned beforehand.

The danger of the above discussion is that language evolution now sounds almost too easy. Bees should not put up with all the fuss and bother of waggle dancing and should instead be holding forth in learned conversation about the finest flowers and the most nourishing nectar. Gorillas should be exchanging dark and disturbing tales about how a group of silverbacks were shot dead in a national park in the Democratic Republic of Congo. Birds should abandon their abstract songs, from the "spring of the yea-ar" sound of the eastern meadowlark to the squeaky wheelbarrow of the black-and-white warbler. They should simply tell it the way it is when it comes to attracting a beautiful mate or driving away a pesky competitor.

That raises the central point, the one we need to understand if we are to figure out what it means to be human: Under what conditions are communicators encouraged to shift from the nonsyntactic communication of animals to the syntactic communication used by humans? Here again, evolutionary game theory can speak volumes. With Joshua Plotkin, who is now at the University of Pennsylvania, and Vincent Jansen of Royal Holloway, University of London, I explored the evolution of syntactic communication, assuming that the players cooperated to achieve common goals.

Most creatures don't elaborate in the remarkable way that we do because syntax comes at a heavy cost. It requires a big power-hungry brain and a degree of mental exertion to string words together in the right way. Natural selection can only see the advantage of this brain expenditure on grammar if the number of relevant communication topics (that affect fitness) are above a certain threshold. If there are only ten things to say to each other ("water," "enemy," "sex?" and so on), there's no point in going through all the fuss and bother of inventing syntax. Having syntax only pays if your ecology includes the complicated politics of indirect reciprocity and there is much to discuss, from boyfriends to iPods to quantitative easing. In a simple environment, where all there is to gossip about is the next banana and the occasional loitering lion, a few grunts and shrieks will do nicely.

In this way, the context and ecology of a life selects the need for language. If we were a great, dim ungulate grazing on grass all day and every day, we would not need to say much more than "yum yum" or "There's nicer grass over there" or "Let's have sex." The same goes for a blue whale cruising through a sea of plankton. Where all there is to gossip about is where the next meal is coming from, how to wield a club, and so on, there is no need for grammar. Actions speak louder than words. But in an environment of sufficient social richness and complexity, where survival depends on important information being quickly dispersed, benefits exceed costs and syntactic communication should win: language should bud and flower. This is especially true if the interactions between members of a group are complicated and politics looms. Alpha males can no longer establish their dominance by brute force alone, but must rely on a gang of cooperative supporters who also want their share. Thus language was born of soap opera and politics.

A LIFE OF SOAP OPERA

There is evidence of fairly complicated animal communication, from insect societies to birdsong, and we only understand aspects of it. There could be the selection pressure toward more sophisticated language in many animal species today, but it seems that the giant leap from animal communication to human language was made only once. Why? Maybe it is a difficult step for neuroanatomy even if the selection pressure is there. Or maybe we do not possess all the necessary information. Maybe there were several origins of language (just as there were several origins of multicellularity). At times there were a handful of human species on earth. Just one—*Homo sapiens*—eliminated all the others.

What is it about our way of life that made language such a necessity? In a nutshell, language was born of our sophisticated society. Among our human ancestors, social interactions got more and more complicated. When there were more opportunities for deception, manipulation, cooperation, conflict—what we call politics—language became a necessity to gain support of others to make deals, to forge alliances, and to collaborate. In turn, language provided still more opportunities to think, reflect, and discuss. Like pouring oil on a fire, language blazed a trail for even more social complexity and hubbub.

In this way, language, brainpower, and society became entwined in a three-way dance. What resulted as each component moved in step with one another was coevolution, a spiral toward more and more social complexity as language allowed for even more manipulation and deception, and ever more collaboration and cooperation too. Thanks to the new invention of widespread indirect reciprocity, coevolution bootstrapped the evolution of the social brain of that remarkable creature, *Homo sapiens*.

The best way to spread the word about my research on language to the extraordinary minds of the institute was by giving talks there. I would inevitably mention the Nash equilibrium. And, quite often, sitting before me in the audience was a grey figure with noticeable ears and a lick of grey hair: none other than John Nash himself, who shared the 1994 economics Nobel Prize for the one-page paper that outlined his influential idea, and whose personal tale of madness and paranoia was popularized in the movie *A Beautiful Mind*. As one would have expected, Nash was an enigmatic figure.

One day, over lunch, I posed a riddle to everyone who sat around me: "The poor have it, the rich need it, it is greater than God, more evil than the devil, and if you eat it you will die." Lindi Wahl and Ramy Arnaout cracked it instantly but did not spoil the fun for the others. The future Nobel Prize–winner Frank Wilczek bounded back after an hour and whispered into my ear, "There's nothing to it." Four hours later, the answer dawned on Karl Sigmund.

Evidently Nash had listened in, as he sat a few yards away. He sent an email the next day, in which he first noted that the riddle was not mathematical but added that a "verbal answer was possible." It continued: "Figuratively one can say the rich need nothing and the poor have nothing and the following religious beliefs are popular: 'nothing is greater than God, nothing more evil than the devil.'" I loved the somewhat prissy, formal way he phrased his answer.

That was the kind of place the institute was. People lived and breathed logic and mathematics. They talked about theory endlessly. They fought over it. They read about it. They dreamed about it too. I once entered Wilczek's office and it looked as if he were napping. But he had heard me step in and immediately stirred, protesting, "I wasn't asleep." He quickly explained: "I was just thinking whether the universe could in fact be five-dimensional and just appear four-dimensional."

IN SEARCH OF GRAMMAR

Two distinct traits of Homo sapiens, *as compared to other species, are technology and social cooperation among non-relatives. It can't be a coincidence that we also have language, the third thing that makes us different.*

—Steven Pinker

What are the conditions that a universal grammar has to fulfill for a population of individuals to evolve a coherent form of communication, one that enables them to understand each other? Or, to put it at its most basic, why is it that babies raised in New York speak American English, and those raised in Amsterdam speak Dutch, and those raised in Vienna speak German?

To find out, my team was joined by an inspiring Russian mathematician, Natalia Komarova, who had done earlier work on nature's patterns, from bubbles and waves to how sand ripples form on a beach. I had first met Natasha at a party and, by the standards of the day, she didn't look much like a mathematician. The stereotype is somewhat geeky and reserved. The old joke says it all: "How do you tell an extrovert mathematician?" Answer: "He stares at your shoes." Natasha, on the other hand, came over like a Russian agent from a cold war spy movie. She was wearing a black leather jacket and, at that time, was smoking. She seemed fascinated by our research and wanted to work with us. She joined our team and over the years would make many important contributions.

We did not always remain in our little house at the edge of the forest when we were working. We often went for walks among the trees. I can still remember one stroll with her when we were discussing the symmetry of a model of the evolution of grammar. She drew the geometry of her latest calculations with a stick in the snow, as lines of black shadow on white crystals of ice. "I don't know the answer yet," she said. "But it will be amazingly beautiful."

At the same time, another remarkable scientist showed up in my office. Partha Niyogi had studied at the Indian Institute of Technology in New Delhi and then done a doctorate in learning theory at the Massachusetts Institute of Technology. At the time, Partha was working at Bell Labs in New Jersey, a powerhouse of research that has been a Nobel Prize factory over its long history. Whenever he came to visit, we would walk together through the Institute Woods, a nature reserve that links a network of green spaces in the state. In the spring, we strolled past yellow trout lilies, pink and white spring beauties, and purple violets. In the summer, we walked through dappled sunlight. In the winter, there was the crunch of snow under foot. He moved on to become a professor at the University of Chicago. A deep thinker, Partha was relentless in his efforts to teach me the mathematical foundations of computer science, formal language theory, and learning theory. Sadly, those efforts came to a premature end. Partha died in 2010 of brain cancer.

Let's examine the simplest recipe for language. We need two people. One is the teacher, the other the student. The teacher utters sentences from his language, selected at random, one after the other. Now
the student must figure out what grammar his teacher is using. After
some time the student must have a (subconscious) representation of
that grammar and be able to produce novel sentences himself. It is
important to remember that the student cannot simply memorize
each sentence, because natural languages, whether English, Bengali, or
Mandarin, have an infinite number of these. To come up with novel
sentences, the learner needs to figure out the rules the teacher is using
to put his sentences together.

The basic question is, what is the set of all grammars that can be
memorized by the human brain? Learning theory shows that this set
of potentially learnable languages must be restricted, so the brain cannot pick up every conceivable grammar. The brain is not a blank slate
that can infer the rules of any grammar. Instead it is restricted to learn
a particular set of grammars. This restricted set of grammars that can
be learned by the human brain is characterized by what Chomsky calls
universal grammar.

For me, an evolutionary biologist, there is a need to move from the
idealized setting of one student and one teacher to the more realistic,
and messy, setting of a community. For evolution, we always need to
think in terms of populations. Thus I wanted to study language learning and evolution in a gathering of speakers and hearers where everyone
says something slightly different. I became intrigued by the question of
how a population of individuals could converge to a common grammar.

In a population there is always a fog of conflicting information.
People might use slightly different grammars. Some folks are cooler
than others and inspire more imitators. Also the evolutionary model
spans many generations of successive speakers. Like the language of
DNA, spoken languages form, mutate, and compete with each other
over many generations. To solve this problem, I extended my work on
evolutionary game theory to language by devising the following model.

Imagine there is a population of people, all trying to talk to each
other. There is a payoff for individuals who successfully communicate

with each other. The payoff improves their fitness, with more genetic or cultural offspring. The biological payoff is that they reproduce faster; they are more efficient at getting mates, for example. In a cultural payoff, they are more likely to pass along their grammar to listeners/learners. Children are not born with a language but with a mechanism for learning language. The critical question is this: What mechanism leads to the evolution of a common language in a population?

When we do the math, we find—unsurprisingly—that when children are presented with too much conflicting information during learning, there is no coherence. Everybody ends up using different "grammatical" rules. The outcome is rather like that seen in the story of the The Tower of Babel in the Book of Genesis.

There are two ways to minimize the errors made in passing grammar on to someone else. First, children need to sample a larger number of sentences. Second, children need a more restricted "search space" of grammars. To put it another way, they need a more specific universal grammar, which is in fact shaped by mutations in genes that determine the basic architecture of the human brain. This is a beautiful coevolutionary process.

One can even work out the threshold at which a Tower of Babel crumbles, along with any chance of our understanding each other. In 2001, Natasha Komarova, Partha Niyogi, and I proposed the idea of a "linguistic coherence threshold" in the journal *Science*. We had found a striking example of an evolutionary "law of nature" that relates the number of sample sentences available to the learning child to the number of candidate grammars in the universal grammar.

Leaving aside the mathematical details, we found that for a given amount of information (number of sample sentences) we could specify how specific a universal grammar had to be for coherent grammar to evolve in a population of speakers. Intuitively, one can see that the more specific the universal grammar, the fewer the sample sentences a child would need for learning a particular grammar. In chapter 6 I described how a similar "law of nature" was found by Manfred Eigen and Peter Schuster in genetics: for a given mutation rate there is a maximum genome length for genetic evolution to be possible. Beyond that, a genetic Tower of Babel collapses to give you meaningless DNA messages.

WORDS IN THE HOURGLASS

A man is well holp up that trusts to you
—Shakespeare,
The Comedy of Errors

Language is in constant flux. The works of William Shakespeare can sometimes be perplexing. Is that odd-looking "holp" an archaic word or a spelling mistake? In turns out that "holp" was the past tense of "help" when Shakespeare wrote these words sometime between 1589 and 1594. It is a telling example because this word is the very stuff of cooperation. As the proverb goes, "If you want happiness for a lifetime—help the next generation."

My work on language would move to the evolution of verbs. However, this phase of my research took place at Harvard, not Princeton. Even though I had financial backing, and enthusiastic support from Harvard, it still took some time to establish my new center there and, in the end, one of my Princeton students would beat me to it, none other than Erez Lieberman, who had spent a year in Israel living beyond the Green Line, the 1949 Armistice lines, on the West Bank.

Erez would often materialize in my office at the Institute for Advanced Study with no warning, and always when I was preoccupied with a difficult problem. But I came to realize that whatever complicated issue was bothering me, I could discuss it with Erez. I love his kind of natural talent, an extraordinary capacity to draw on unconventional and ingenious methods, while Erez in turn saw me as a "source of good problems."

Erez did two student theses, one in math and one in philosophy. In the latter he focused on the philosophy of language outlined by the great Wittgenstein and interpreted by the distinguished modern philosopher Saul Kripke. The possibility of our ever following rules in our use of language is apparently undermined by that philosophy, which is jokingly dubbed "Kripkenstein" to reflect how it is Kripke's particular and controversial interpretation of Wittgenstein. Erez won first prize for the thesis he did with me in math on learning finite languages, which turned out to be a relative of the classic sphere packing problem. To be learnable, languages must be distinct enough that they do not overlap, rather like the oranges in a greengrocer's crate.

Erez told me he wanted to apply to Harvard graduate school, after doing one year of rabbinical school in New York. Before I had any plans to go to Harvard myself, I wrote an enthusiastic letter of recommendation for him. I soon found myself answering a phone call. The voice at the other end told me that Harvard University was "considering Erez Lieberman's application." Lieberman's resume looks good, the bureaucrat explained, but no letters of reference had arrived. Realizing that my letter must have gone astray, I replied that Erez was simply amazing. He was accepted. In this way, Erez had beaten the record set by the famous beautiful-mind of game theory, John Nash, who was accepted by Princeton University with the shortest letter of reference ever. It simply declared: "This man is a genius." Lieberman apparently was accepted by Harvard without any letter of reference at all.

The next year Erez became my graduate student at Harvard. He made quite an impression when he declared his philosophy: Typically people invest their whole life studying the few things they learn in their doctoral thesis. But Erez did not want to be pigeonholed this way. He wanted to use his thesis as a vehicle to learn everything and anything. For him, a thesis was not a taster of things to come but an intellectual *Smörgåsbord*. Erez has already finished enough projects for three or even four PhD theses and has only just got around to defending his first.

He was intellectually promiscuous, collaborating with anyone who would listen to him. Before long, he began research with the great genome tsar Eric Lander. When we visited the Google offices in Cambridge, we all had to sign in. Not Erez, however. He was already well known there and could come and go at his leisure.

Then, for tragic reasons, his research agenda changed again. His grandmother had a serious fall, which was fatal, so Erez decided to work on the balance problems of astronauts with the National Aeronautics and Space Administration. That led to a start-up company that made a smart sole for a shoe to diagnose how badly someone is teetering. The algorithm he used to analyze signals from the sole is related to one used by Lander to find genes.

Finally Erez became involved in our ongoing research on language evolution. Kripkenstein's point about the rules of language was, he said, "still percolating in my head" when one day, it was given new salience by a talk on irregular verbs by Steve Pinker, the distinguished Harvard language researcher. The general rule in English is that you conjugate the past tense of verbs by adding -ed but there are exceptions. So many in fact, that Erez became intrigued. How is it that people can learn to disobey the rules of language for this laundry list of exceptions? How does this tendency change with time? Working with the creative French doctoral student Jean-Baptiste "JB" Michel and the "Romans" Joe Jackson and Tina Tang, Erez and I decided to investigate. It took several years to come up with the data for the study. In the end, we focused on the evolution of verbs. Our mission was simple: we wanted to predict the future of the past tense.

Thanks to the heroic efforts of Tina Tang, who scoured texts for examples, we were able to study the evolution of 177 irregular verbs over the past 1,200 years, from *Beowulf* to *The Canterbury Tales* to *Harry Potter*. We grasped something that no one thought could be measured and obtained striking results. Of the seven rules for conjugating the past tense in Old English, only one has survived. That single extant rule adds an -ed suffix to simple past and past participle forms. Just as genes and organisms undergo natural selection, words—specifically, irregular verbs such as "holp" that do not take an -ed ending in the past tense—are subject to powerful pressure to "regularize" as the language develops.

Of the 177 verbs that were irregular 1,200 years ago, 145 stayed irregular in Middle English. Only 98 or so remain irregular today, following the regularization over the centuries of such verbs as "help," "laugh," "reach," "walk," and "work." What's striking is that the "decay of an irregular word" followed a very clear trend with remarkably small error bars. The mathematical function explaining this decay says that a verb used 100 times less frequently will regularize 10 times as fast. To put it another way, verbs evolve at a rate inversely proportional to the square-root of their prevalence in the English language. An irregular verb that occurs less often will be forgotten faster. In this way, irregular verbs behave in just the same way as radioactive atoms and have half-lives. We can calculate the half-life of irregular verbs depending on their frequency.

The present tense of "I *know*" has the irregular past tense "I *knew*." Although little kids might apply the logic of language to say—quite understandably—"I *knowed*," this verb hasn't regularized yet. "I *know*" is so common that it resists change. Children also start out by saying "growed" before they learn the -*ew* rule. Or "hitted" instead of "hit"— the basis of that old joke—Child: "Mummy, Bobby hitted me. Mother: "Bobby hit me." Child: "You too? Boy, that kid is in trouble."

Children need to hear the irregular form often enough to memorize it. As a corollary of this, relatively less popular words are more likely to succumb to change. And all modern verbs, such as "to google," are regular. Overall, the mathematics reaches a conclusion that would appall the diehards of the Académie française who are still defending the purity of the French language from ghastly English imports, from "*le parking*" to, horror of horrors, "*le washing up*." They are probably wasting their time. You can't defy evolution. By the same token, the defenders of the Queen's English are adopting a Canute-like pose.

THE FUTURE OF CULTURE

Tina Tang faced a huge chore in having to go through academic texts to track the evolution of English verbs. If only we had waited for a few years. Her efforts were soon overtaken by technology. The Google Books project has digitized a vast range of tomes, including many of the texts that we had laboriously scoured. Erez realized that we could create a tool with Google to mine this sea of literary data for nuggets of information. Yuan Shen, an MIT graduate student, and Jean-Baptiste Michel worked with Erez and Google to come up with Google Bookworm, a tool that allows us to track major trends in language based on a collection of millions of books, and about 500 billion words in all. That is enough to reveal not just linguistic patterns but cultural ones too.

We can search all half trillion words in the database to reveal the trends across geography, time, and whatever. Erez was particularly taken by one plot showing the use of the phrases "World War 1" and "World War 2" and that alternative name for the earlier 1914–18 conflict, "the Great War." That latter was used to signal that it seemed huge, mechanized, and quite unlike any before. But one can see the phrase "the Great War" decline during the 1940s with the realization it was the first of two great global conflicts.

One could plot those rare occasions that regular verbs become irregular, such as the rise of "snuck" after the millennium. You can chart the rise and fall of swear words, finding an intriguing link between expletives and the rise in violence. There's plenty of evidence that the word "evolution" is ruled by the "rich get richer" phenomenon, where common terms become even more so, while rarer terms fade away.

Erez and JB created the biggest searchable language data set ever from about 5 million books (in various languages, including English, German, Russian, and Chinese). Erez likes to put it like this: "If you write out the entire text in normal font size, you can get from the Googleplex in Mountain View to the moon and back twenty times. In comparison, if you write out the second biggest database, the British national corpus, you only get from Mountain View to the launching pad in Cape Canaveral."

This language database is a remarkable resource. One can use it to get the measure of culture and the way it changes and evolves. Erez likes to call this endeavor "culturomics." By studying the "culturome," one can discern the pulses and rhythms of society, from the ebb and flow of epidemics to the rise and fall of technologies. By scanning the database for 154 inventions created between 1800 and 1960, from microwave ovens to electroencephalographs, we found that the more recent innovations took far less time to become widely adopted. We found that God is not dead but is in need of a new publicist. Due to the rise of secularization, one can see "God" and "Jesus" have slowly declined in use over the last two centuries, though they still trump every other name by a wide margin (however, for a few years in the German corpus, "Hitler" was more frequent than "Jesus"). We were able to study the impact of censorship and suppression. During the Third Reich, the names of artists, writers, political academics, philosophers, and historians all began to vanish from German texts, while the names of Nazi party members flourished to become six times more common. And we could see the fickle nature of fame. It was easy to chart the way people rise to become a household name, only to fade and disappear. Unfortunately, politicians trump all other professions in terms of absolute fame. Typically we are talking about American presidents, or the likes of Napoleon, Hitler, and Churchill. Actors reach the height of fame younger than others. Alas, science is a relatively poor way to achieve wider recognition, and mathematicians are the least famous of all the different flavors of scientist. The culturome of celebrity revealed that the best-known people alive today are more famous—in books at least—than their predecessors. However, this fame is increasingly short-lived: today's stars shine brighter and burn out faster.

THE LANGUAGE INSTINCT

Overall, my research supports the idea of a universal grammar that was first put forward by Chomsky. Only with this idea can we understand how humans are able to learn different languages. Today these number some six thousand in all. Which one you speak of course depends on where you are raised and by whom. It could be Tojolab'al, used in the Chiapas region of Mexico, Chiwere, a Siouan language spoken by the Otoe-Missouria and Iowa tribes in the United States, or, in Australia, the aboriginal languages of Magati Ke, Yawuru, and Amurdag. If, that is, anyone is left to speak them. Some fear that half of all languages may go extinct within a century.

I have also shown that the evolution of cooperation is tightly linked with the evolution of language. There had to be selection pressure for the evolution of language, which means that it boosted the fitness of ancestors who had the gift of the gab. For example, language may have made us better hunters, though we know from other creatures that the frustration of hunting without being able to brag about it is in itself not enough to drive the evolution of our complex form of communication. More important, the emergence of language led to an extraordinary range of social interactions and discussions about who did what to whom and why. As a result of language, social life became ever more sophisticated. Our brains did too. Big brains are costly in terms of food and childbirth, and the development of our extraordinary encephalon could only be justified by making us adept at the complex social politics that come with a sophisticated language. Down the generations, as human society expanded and became ever more elaborate, the gossip that lubricated the mechanism of indirect reciprocity also made us smart.

CHAPTER 10

Public Goods

*Human activities have already demonstrably changed global climate,
and further, much greater, changes must be expected throughout this
century. The emissions of CO_2 and other greenhouse gases will further
accelerate global warming . . . Some future climatic consequences of
human-induced CO_2 emissions, for example some warming and sea-
level rise, cannot be prevented, and human societies will have to adapt
to these changes. Other consequences can perhaps be prevented by
reducing CO_2 emissions. Everyday measures can contribute to climate
protection.*

—Jochem Marotzke

Anyone who has been following the gloomy headlines about climate
change will find little of novelty and surprise in this quotation from
an advertisement in the German newspaper *Hamburger Abendblatt*.
Even though its message is alarmist, it is now so familiar that it would
provoke little comment. But there is a fascinating story behind this
particular ad, one that gives us some crucial insights into how people
can cooperate to save planet Earth.

The copy was not written by an advertising agency but by the director of the Max Planck Institute for Meteorology in Hamburg, Jochem
Marotzke. Nor was it paid for by a big company that was keen to
show off its eco-credentials. Nor even a jet-setting philanthropist who
wanted to buy green credibility. The money came from the proceeds of

a special experiment devised by Manfred Milinski, of the Max Planck Institute for Evolutionary Biology, a connoisseur of food and fine wine, a zoologist and naturalist who feels equally at home with people and stickleback fish, as we saw in chapter 1. At the heart of the experiment was an ingenious game that Milinski had devised to examine the extent to which we can all pull together in a crisis.

What is remarkable about Milinski's game is that the real-life version is played by around 7 billion people every day of every week. It has come to dominate their lives. Yet many players still seem blithely unaware that they are involved, let alone that this global game has a name. This huge undertaking can be thought of as a variant on the Prisoner's Dilemma, known as a public goods game.

The Prisoner's Dilemma is a game for two people. Any more than this, and you end up with a public goods game. On planet Earth, of course, there are billions of players. If one person defects, harming the environment and thus the interests of others, then a common way for fellow players to retaliate is to defect as well. So, for example, if we see litter on a street we are less likely to worry about adding a little litter of our own. This pattern ends up hurting everybody.

Nowhere is the impact of this vast game more acute than on the global environment, where selfish motives to consume or pollute seem to outweigh the collective good, day after day, week after week, year after year. Fortunately, one of the mechanisms of cooperation can play a central role in solving this exotic version of the Prisoner's Dilemma. And that mechanism is indirect reciprocity.

The good news is that the Climate game that paid for Marotzke's advertisement illustrates how people can sometimes pool their efforts when the stakes are high and when the power of reputation is harnessed to catalyze cooperation. The bad news is that after running a number of experiments, Milinski found that failure is more common than success. More depressing still is that such games suggest that when politicians get involved, the outcome is gloomier, though not by much. Politicians don't seem to have much impact. Certainly less than they would have us all believe.

THE TRAGEDY

The best-known public goods game of all is called the Tragedy of the Commons. The Tragedy was first referred to by Garrett Hardin in his eponymous 1968 paper that appeared in the journal *Science*. The paper proved highly influential and, being more relevant today than ever before, has recently been reprinted dozens of times. Hardin had illustrated the Tragedy with the example of a "pasture open to all" that herds of economists, social scientists, and game theorists have grazed on ever since. The phrase has become the best-loved metaphor among experts to illustrate our chronic inability to sustain a resource that everybody is free to use and, alas, is just as free to abuse.

In his paper, Hardin described how livestock owners allowed land that was held in common to be overgrazed, even though they realized that by doing so the commons was diminished to everyone's detriment, including their own. They grazed their own animals there, worrying that if they refrained others would take advantage of the resource and it would be trashed without any benefit to themselves. The "tragedy" arises when, to make sure they get a share, each herdsman adds cows to a common field despite the costs of overgrazing to the whole community. It is easy to imagine this chain of reasoning: "Should I add another cow to that common meadow? What is the advantage for me? What is the downside for everyone? If I graze a couple more head of cattle, it will hardly make any difference. Besides, that other guy has many more cows on the common pasture than I do."

But, of course, everyone thinks the same way. There's no incentive for herdsmen to be cautious. Those who graze more cattle have a larger net benefit than those who hold back, worrying about the fate of the pasture. The cattle gradually increase to the point where, as more are added, their peers get thinner because there's not enough food to go around.

The tragedy comes in various flavors. In Hardin's classic paper the root of the problem is overuse. Grass can only replenish at a certain rate so the pasture is ruined if overgrazed by enough cattle. This idea of excessive use could apply equally well to other limited resources, from oil to furry animals. Or fish, for that matter. Back then, Hardin himself issued a warning about the world's oceans that has particular resonance today: "Maritime nations still respond automatically to the shibboleth of the 'freedom of the seas.' Professing to believe in the 'inexhaustible resources of the oceans,' they bring species after species of fish and whales closer to extinction." Remember that his prophetic words were published back in 1968. More than three decades later, a study warned that 90 percent of large fish have disappeared in the past half century and said that the bigger fish—large tuna, sharks, swordfish, and even cod—may soon be memories.

Hardin later gave a capsule summary of his landmark paper: "In a crowded world, an unmanaged commons cannot possibly work." That is an important qualification. If the world is not crowded, a commons may in fact be the best method of distribution. For example, he explained, when the pioneers spread out across the United States, the most efficient way was to treat all the game in the wild as an unmanaged commons ("just fire away") because for a long time humans couldn't do any real damage. "A plainsman could kill an American bison, cut out only the tongue for his dinner, and discard the rest of the animal." He was not being wasteful "in any important sense," says Hardin. Nor did it much matter how a lonely American frontiersman disposed of his waste. Today, with only a few thousand bison left, we would be appalled by such careless behavior. As the population in the United States became denser, the land's natural chemical and biological recycling processes were overloaded. Careful management of these resources became necessary, from bison to oil and water.

The tragedy includes the equally damaging corollary of overexploitation: adding something harmful to an environment. Now the tragedy reappears in the guise of pollution. "Here it is not a question of taking something out of the commons, but of putting something in—sewage, or chemical, radioactive, and heat wastes into water; noxious and dangerous fumes into the air; and distracting and unpleasant advertising signs into the line of sight," said Hardin.

Exactly the same thinking can be applied to underinvestment in a common resource, such as failing to weed a garden, fix transport infrastructure, or even lay a communal carpet in a house. The problem could be couched either as "overuse" of the carpet by people with muddy shoes or as "underinvestment" in shoe cleaning and vacuum cleaners, for example. And there are endless examples to be found in the financial industry.

Here's one. Many people use companies to invest in the stock market on their behalf in the form of a fund. There is a basic choice on offer. In an active fund, a fund manager will pick shares, depending on the performance of companies, their potential, and so on. Because active funds depend on research, they are more expensive to run than the alternative, a passive fund, which tracks the stock market average by investing in a portfolio of major companies. Because passive funds deliver an average performance, half of active funds will do worse than the passive funds. Half will do better. But because there are thousands of active funds to choose from, the easiest and most rational approach for most people is to choose a passive fund. It is safer for an inexperienced investor and there is less overhead. If everyone did this, though, there would be little research carried out on the potential of companies. Starved of intelligence, the market would break down.

Many examples are also to be found in the digital pastures of the internet. Common resources—from free software, such as GNU, to eBay and Craigslist—can benefit many people but are also open to exploitation by defectors and cheats. As we have already seen, Google page rankings, the reputation of eBay buyers and sellers, and the Amazon.com reader review system are all based on trust.

CRUELTY AND KILLING

Hardin had highlighted a general class of problem, one that became more pressing as the planet became more populous. But he was the first to admit that this insight was by no means new. It was as if Hardin was exasperated that the collective penny had taken so long to drop. He pointed out that the rebuttal to the idea that Adam Smith's invisible hand could control population could be found long ago, in "a little-known pamphlet" in 1833 by a mathematical amateur named William Forster Lloyd (1794–1852).

Hardin was concerned that little progress would be made in dealing with the population problem "until we explicitly exorcize the spirit of Adam Smith." He was referring to Smith's magnum opus, *An Inquiry into the Nature and Causes of the Wealth of Nations,* which appeared in 1776, at the dawn of the Industrial Revolution. In this influential work, the Scottish moral philosopher advocated a free market economy and popularized the idea that an individual who "intends only his own gain," is, as it were, "led by an invisible hand to promote . . . the public interest."

Hardin saw only ruin in this idea. Smith claimed that individuals acting in their self-interest will act to increase the common wealth. Hardin countered that self-interest will ruin collective wealth and cited the fourth American president, James Madison, who said in 1788, "If men were angels, no Government would be necessary." Smith's claim is only true, explained Hardin, if all men were angels. As my work has shown again and again, anywhere you can find angels, demons of defection also lurk. "In a world in which all resources are limited, a single non-angel in the commons spoils the environment for all."

Born 1915 in Dallas, Texas, Hardin's formative years would plant the seed for his later, influential ideas, which also included his thinking on what he called "lifeboat ethics." ("Ours is a limited world, and we have to find out how to dispense the goodies.") At the age of four, Hardin was afflicted with polio—a virus that causes paralysis. After being bedridden for weeks, he was not strong enough to walk very far when he at last started first grade. Children can be cruel. At school, Hardin was labeled as a cripple by his little peers. The ridicule may have driven him on to excel academically.

Although his family moved quite often, a farm located a few miles from Butler, Missouri, remained a fixed feature of his childhood. From the age of ten and through his peripatetic teenage years, he spent each summer on the farm. Early on, at the age of eleven, he was responsible for feeding hundreds of chickens and killing one each day for lunch.

He would come to say that learning to kill an animal is an important part of everybody's education.

Life on the farm would be relevant not only for his emerging views on survival but on the environment too, thanks to the misguided kindness of strangers. The local cat population was fueled by city people dropping off unwanted cats. They thought that it was kinder to abandon their felines in the country than kill them. There, of course, the farm dogs would attack them. Or they would have to be culled. Or, when the population grew too big, cat flu would sweep through and take its toll.

In the farmyard, Hardin learned firsthand that death is part of life. During his upbringing he obtained visceral insights into the population problem. For the rest of his life he carried a strong conviction that there is a limit to the carrying capacity of an ecosystem. "I have been haunted by the realization that there simply isn't room for all the life that can be generated, and the people who refuse to cut down on the excess population of anything are not being kind; they are being cruel. They are increasing the suffering in the world."

For Hardin, this was not a matter reserved for the ivory tower and for dry, academic debate. He practiced what he preached. He and his wife, Jane, were both members of the Hemlock Society, which campaigned for voluntary euthanasia. On September 14, 2003, Hardin and his wife died at their home in Santa Barbara, California. He was eighty-eight and she was eighty-one. Both were in poor health. They had committed suicide shortly after their sixty-second wedding anniversary.

THE POPULATION PROBLEM

While Hardin was at the University of Chicago, another key influence would shape his thinking, in the form of W. C. Allee, one of the early ecologists who, despite the relatively low birth rate of his day, would warn of the perils of overpopulation. According to Hardin, his professor would mutter that "it's bound to happen, it's just a matter of time when things improve then zip, up it'll go."

Some like to argue that the carrying capacity of the world is only about half what the population is today (and that is being generous when it comes to numbers of people, since it assumes only a modest lifestyle). Hardin's take on the problem seems prescient: "To couple the concept of freedom to breed with the belief that everyone born has an equal right to the commons is to lock the world into a tragic course of action."

"Freedom to breed will bring ruin to all. At the moment, to avoid hard decisions many of us are tempted to propagandize for conscience and responsible parenthood. The temptation must be resisted, because an appeal to independently acting consciences selects for the disappearance of all conscience in the long run, and an increase in anxiety in the short. The only way we can preserve and nurture other and more precious freedoms is by relinquishing the freedom to breed and that very soon."

Hardin, who became a professor of biology at the University of California, Santa Barbara, realized that there was a fundamental flaw in the thinking of his day. "Most people who anguish over the population problem are trying to find a way to avoid the evils of overpopulation without relinquishing any of the privileges they now enjoy." Then, as now, many people were seduced by the idea of a high-tech fix, one that would mean they did not have to act. "They think that farming the seas or developing new strains of wheat will solve the problem—technologically," he said. Today, for example, the elusive goal of commercial fusion power is one oft-cited technical fix to global energy problems. But in his paper of 1968 he concludes: "I try to show here that the solution they seek cannot be found."

The key to solving the Tragedy of the Commons was nothing to do with technology. It required what he called a "fundamental extension in morality." What did Hardin mean by this? Let's take, as an example, the per capita carbon dioxide emissions of the United States, which are twice as high as those of the United Kingdom and three times as high as the emissions of France or Sweden. The reason for this amazing disparity cannot be that the United States lacks the necessary technology or wherewithal to devise a fix, given its extraordinary dominance in research and innovation.

The answer to this riddle is that many Americans are unwilling to change their behavior and to relinquish gas-guzzling cars and a high-energy lifestyle. Many seemingly do not see it as immoral to waste and poison the environment. They just "drive on." I agree with Hardin that the Tragedy of the Commons has no real technological solution, only one rooted in ethics and behavior. In short, we must enhance the way that we cooperate with each other on a global scale. Good planets are hard to find.

Look up at the night sky on a moonless night, far away from the orange sodium haze of street lights, and marvel at the light show. You can commune with your ancestors this way. Ancient mariners would gaze in wonder at the river of light that is the Milky Way, our home galaxy. That wondrous canopy of stars has inspired poets, philosophers, and dreamers for millennia. Given what we learned in chapter 6, some of those twinkling points of light may well have planets that harbor life. Hardin's conclusions are so universal that one cannot help but wonder how often alien intelligent life has already been snuffed out because it was unable to solve the Tragedy of the Commons.

THE PUBLIC GOODS GAME

Without the intervention of a third party, such as a government, Hardin thought that selfish interests would come to dominate the commons in a "destructive" way. And, with the help of a standard experiment, it's easy to see why this would happen. Four people are each given an endowment of $8 and told that they can choose to invest between nothing and $8 in a pool by anonymously putting the money in an envelope. The experimenter collects the contributions, totals them up, adds half again to the amount (in other words, multiplies by 1.5—this profit would be the equivalent of the profit made by the herdsman when he finally sold the cattle he has grazed on the commons), and then divides this money evenly among all the players.

If all four people contribute their $8 to this public goods game, the initial pool will contain $32. Multiply that by 1.5 and you end up with $48. Divided equally among the four players, each ends up with $12. Thus, by cooperating, they have all made a profit of $4. But there's a problem. What if one player refuses to put any money in the pot and hangs on to his $8? The others still invest $8 each so that $24 is in the pot. Multiply $24 by 1.5 and you have $36. Divide equally among all four players and they each end up with $9. But while three of the players have made a profit of a dollar, one player has made $9 by contributing nothing.

Place yourself in the position of a player and you can see the dismal logic of defection. Most people are optimistic initially. You'll invest your $8 and you should make $12—if everyone does the same. But, of course, there might always be one who holds back on investing and takes advantage of others' generosity. When the return on your investment is less than $12, you will realize that someone has not invested her fair share and be tempted to hold back as well. There is a strong incentive to let everyone else contribute so that you can reap the benefits of their investments. Now if everyone thinks the same way, nobody invests and nobody makes any profit. The rational course is to invest nothing. This is the Tragedy of the Commons. One can also think of this public goods game as a Prisoner's Dilemma where more than two people play simultaneously.

The essence of this Dilemma, and its relevance to climate change, was captured by the games that were used to pay for the advertisement in *Hamburger Abendblatt* that I mentioned at the start of this chapter. In all, the games were played by 156 undergraduates from Hamburg University, in a computerized experiment conducted by Manfred Milinski. The players were divided into twenty-six groups of six subjects. This time the aim was to see whether they would contribute their own money to sustain the global climate in a public goods game.

Unlike traditional variants of this game, the contents of the public pool were not redistributed among all players but transferred to what Milinski called a "climate account," after the students had made their contributions and the total amount had been doubled by the experimenters. The students were assured that the money from the climate account would be used to pay for an advertisement in a widely distributed daily newspaper. The size of the advertisement, or the impact of its message on the public, would depend on how much money they earned.

They found that players can behave altruistically to maintain the Earth's climate given the right set of circumstances. The first ingredient of cooperation was information. The students were more altruistic when provided with expert information describing the current state of knowledge in climate research. Furthermore, if players were allowed to make their contributions publicly instead of anonymously, personal investments in climate protection increased substantially. The reputation effect was surprisingly strong, according to Milinski. People really do like to be seen to do the right thing.

THE CLIMATE GAME

In an elegant experiment, Milinski and his team came up with another insight into what motivates people to give the commons due respect. This time the game was conducted over ten rounds with six players to show how to deal with dangerous climate change. They organized the game to study whether a group can reach a collective target through individual sacrifices, when everybody is sure to suffer if they don't achieve their target.

This game scenario is most realistic if we allow levels of greenhouse gases to continue to rise at the current rate. Many of the extrapolations assume a smooth rise in the risk of mishap as levels of carbon dioxide rise. But there are some, such as the father of Gaia theory, James Lovelock, who have expressed concerns about abrupt change if the climate crosses certain thresholds and undergoes fast, irreversible transitions. For example, the deep circulation in the Atlantic Ocean could collapse, switching off the warmer currents that help keep winters mild in Britain.

In an extreme case of a climate flip, the temperature in northwestern Europe could fall up to 5 degrees Celsius. A version of this scenario was popularized in Roland Emmerich's film *The Day After Tomorrow,* where a change in ocean current circulation caused by global warming triggers snowstorms in New Delhi, tornadoes in Los Angeles, and ice sheets that move faster than a man can run. This is wildly over the top. In reality, such changes would occur over decades. Even so, by the standards of evolution and geologic history it is an eyeblink. Even if these climate lurches never come to pass, some societies are highly vulnerable to even modest levels of climate change. Poor nations and communities are particularly at risk of disruption, from major migrations to wars over precious resources, such as water.

Against this vivid backdrop of what losing the Climate game really means, all players start out with 40 euros in their private accounts. In each round, players can transfer 0, 2, or 4 euros into a "climate account." One can think of their investment as being equivalent to giving up flying, leaving the car at home to walk to the local grocery store, or abandoning other activities that drive climate change. Note that this particular Climate game is not identical to a classical public goods game but a variant on that theme with different rules. In the former, there is no incentive to give anything (as game theorists say, the Nash equilibrium is to give nothing). In Milinski's game, if everyone gives exactly 2 euros in each round then there is no incentive to deviate from this.

The players were told that after ten rounds the game would end and a computer would tally up the climate account. The team would win if the climate account contained at least 120 euros. That would mean, given there are six players, that each one had to contribute an average of 2 euros per round. If they did this, the money they'd contributed would be the equivalent of bringing carbon dioxide emissions to safe levels, and thus saving the world. As a bonus, each player received whatever is left in his private account, which works out at 20 euros. In the simplest variant of the game, losing means that nobody gets anything. They go home empty-handed. However, at least they have a home to go to. (If we lose the real climate game that may not be the case.) But, of course, they can invest nothing and hope others make up the shortfall. After each round, they would be told how much was invested, but if the tally suggested someone had not paid his fair share, no one would be able to figure out who had shortchanged the climate account.

To make the game easier to play, and to analyze, the players had three choices. Milinski classified the investments of 0, 2, or 4 euros as "selfish," "fair," and "altruistic." Stepping back from the game for a moment, it is easy to translate this into something relevant to the climate debate.

At the time—pre Barack Obama's more enlightened policy on climate change—that the paper came out, the United States would count as "free riders" who did the equivalent of contributing nothing, the United Kingdom played "fair," and France and Sweden would be "altruists."

So if all players always play fair, then the climate account will reach exactly 120 euros, the climate will be saved, and every player will keep 20 euros in his private account. Note that if one player contributes more, then he will have less income in the end. If one player contributes less, then the target will not be reached and the expected income for all will be much lower. This is an example of a Nash equilibrium, wherein players look for outcomes in which each player is making an optimal choice, given the choices the other players are making.

People, however, may not stick to the Nash solution. There is an incentive to contribute less and hope that others will, in turn, compensate. If one player invests 0 in one round and is a free rider, then another player must be altruistic and invest 4 for the total sum to stay on target. This aspect of the game adds a little twist: the free riders now have to rely on the altruists to save the climate. They think that by giving nothing they can force others to donate. But without the altruists, the free riders risk losing their money too. So we come to the conclusion that without altruists there is no incentive to free ride. Without saints, there are no sinners.

To take into account the uncertainties of real life—and there are indeed many when it comes to climate change, including "unknown unknowns"—the game came in three versions, where losing led to a 10, 50, and 90 percent chance of disaster. In the latter, if players failed to invest 120 euros overall, Milinski devised the game so that there could be a 90 percent chance that the climate is lost and thus a 90 percent chance that all the players lose all of their money—the money in the climate account and in their personal accounts too. That meant that there was a one in ten chance that the players could take their money home even if the climate account did not reach its target.

When the game was played by ten groups of six students with the 90 percent chance of disaster, half succeeded. Those groups who failed had accumulated 113 euros on average in their climate account after ten rounds. Ironically, some of the groups came very close to the target but fell slightly short. Players tended to experiment with gamesmanship at first, so that the group loses more and more ground. By the final rounds, there's usually little they can do to recover from the deficit.

Here is a typical example of how the games played out. After eight rounds of one particular game, the climate account contained 90 euros. In the ninth round, to save the world, four of the six players contributed the maximum amount of 4 euros each. The two remaining players were free riders. In the final round, one of the free riders contributed 2 euros while the other remained stingy. Three of the altruists gave 2 euros each. They needed 14 euros but only came up with 8. It seems that the altruists felt that they had already contributed enough. The motives of the free riders were unclear. The final amount came to 114 euros. Everything was lost.

What happened when the link between playing the game and wrecking the planet was more remote? In one version there was a 50 percent chance that the climate would be lost if the target sum was not reached (where hanging on to your money pays out as well as the fair strategy). In the second version, designed to encourage even riskier behavior, there was only a 10 percent chance the world is doomed (when the rational strategy is to hang on to your 40 euros, since this pays out 36 euros over ten experiments compared with the fair strategy, which only yields 20).

What happened this time? Milinski found that only one of ten groups reached the target in the 50 percent version and not a single one of ten groups succeeded in saving the world in the 10 percent version. This outcome is not surprising because in both cases there is no rational incentive to invest in the climate account. In fact, it is astonishing that in these circumstances people invested any money at all in saving the world. Yet in the 50 percent and 10 percent treatments, people donated on average 92 euros and 73 euros, respectively. These investments may have been the consequence of a "framing effect"—participants were told that the game was about saving the climate and thus the world. The conclusion is heartening in one sense: people are willing to gamble for the climate. But the findings are depressing in another: unless they fully realize the extent to which the planet is in peril, people will fail to do enough to save it.

GAME THEORY CAN SAVE THE WORLD

Much effort has been invested in trying to work out how to protect commons. Elinor Ostrom, who is affiliated with Indiana and Arizona State universities, has looked at the role of sanctions in tending commons, whether fish stocks or pastures—what she calls common-pool resources, or CPRs. Based on real-world evidence she amassed on the management of common pools, she concluded more tolerable outcomes arise when users themselves devise rules and enforcement mechanisms. But she concluded that sanctions should be graduated, mild for a first violation and stricter as violations are repeated. Her pioneering insights into how to resolve conflicts led to her sharing the Nobel Prize in Economics in 2009.

In this chapter, we have shown an alternative to the use of punishment and sanctions. One fundamental conclusion drawn by Milinski and his team is that the public must be well informed about the risks of climate change. Ordinary people must have a reasonable understanding of what is going on with the global ecosystem. If the public is misled into thinking that the risk is small, then they will not cooperate. If people know that the risk is high, then they will be much more inclined to club together to curb climate change.

The role of scientists must be to provide honest, reliable information. If they embellish and inflate the risk, then there is a danger that they will lose the confidence of the public. To cry wolf can turn out to be as damaging as underplaying the risks. There are many who feel that the dangers of BSE ("mad cow" disease), AIDS, and swine flu were exaggerated (and, of course, there are many experts who rightly counter these arguments by pointing out that the death tolls would have been much worse if they had underplayed the risks). Like some other highly charged aspects of science, such as embryonic stem cell research, germ line gene therapy, and conservation, passionate advocates must take care not to spin and distort, even if they mean to back a good cause. They must accept the results of good quality research and peer-reviewed studies, even if they undermine their beliefs. They must focus on the positive effects of climate change as much as the negative.

There is a related issue: public understanding of science. Many climate change predictions are couched in terms of risks and probabilities. They rest on making certain assumptions. When presenting this information to a public that is hazy about the difference between climate and weather, or finds it hard to work out a percentage, even a clear, carefully drafted message can be misinterpreted. There is evidence in Britain, for example, that careless presentation of seasonal forecasts harmed public confidence in the predictions.

Transmitting the message with high fidelity is crucial. As Hardin realized, although we must invent environmental solutions, from wind power to fusion energy, only behavioral solutions can save us in the long run. We must learn how to cooperate on a global scale, to respect the needs of others, and to avoid an excessively wasteful lifestyle, where everyone "just fires away," as Hardin put it. Today, we need to avoid a culture where everyone "just drives away."

One way that we can become more familiar with the Tragedy of the Commons is for us all to play the kinds of games devised by Milinski. Let's do it at company retreats, at schools, and in the home. Let's devise a fun version for the web. We all need to get the feel for being involved in a global-scaled "collective-risk social dilemma" and learn strategies for its solution.

Cynics may sneer at the prospect of applying the findings of idealized experiments to the real world. Admittedly, the scale of the real thing is daunting. The group playing this climate "game" consists of 7 billion individuals. The real climate game does not consist of rounds. No one knows how well we are doing when it comes to curbing carbon dioxide emissions. And, indeed, experiments conducted by Milinski on public goods games suggest that the more players there are, the harder it is to cooperate.

In this respect at least, there seems to be a ray of hope: all the big decisions are made by relativity small groups of politicians, such as the G8 leaders, the heads of the Group of Eight forum, who represent the governments of eight nations of the northern hemisphere: Canada, France, Germany, Italy, Japan, Russia, the United Kingdom, and the United States. Perhaps this small number improves our chance of cooperation. And because they are not relatively naïve biology students, but sophisticated well-advised politicians, perhaps the outlook is even rosier. Milinski has done experiments to investigate this theory, but alas, placing the fate of the Earth in the hands of a few politicians does not seem to make much difference. He explains: "The politicians lost out in our games because people wanted them to invest less than other politicians did for their countries. Those who invested their country's money to help rescue the climate lost their reputation within their country."

However, let's return to a point raised by the game that generated the advertisement. The players were more cooperative if their peers could see how generous they were. It sounds glib, but reputation is a very powerful force. In fact, it is much more powerful force than many of us realize, one that has been harnessed across human societies for millennia.

THE POWER OF REPUTATION

On each landing, opposite the lift-shaft, the poster with the enormous face gazed from the wall. It was one of those pictures which are so contrived that the eyes follow you about when you move. BIG BROTHER IS WATCHING YOU, the caption beneath it ran.

—George Orwell, *Nineteen Eighty-Four*

Totem poles are monuments to the power of reputation. They are erected for various reasons, from the mortuary poles raised in honor of a person who has perished, to memorial poles that commemorate important occasions. Some of the decorations are recognizable, from frog to beaver, raven, wolf, bear, eagle, and human; others are more mysterious, varying hugely from family to family, clan to clan, and place to place in the Pacific Northwest of North America. The faces on the poles can be dramatic, with open mouth and bared teeth. They are vigilant, with alert, black painted eyes that seem to miss nothing. The eyes are honed from cedar, yet so sensitive are we to the power of reputation that these wide eyes have an effect on us. The decision to paint eyes that seem to see members of a tribe exploits the fact that the more people know that they are being watched, the more charitable they become. Cooperation kindled by indirect reciprocity has led to an arms race when it comes to establishing one's own reputation and discerning the reputations of others.

No wonder that George Orwell's Big Brother, the dictator of Oceania, was always watching the citizens of the totalitarian state, or that religions contain the idea of an omnipresent God who "sees through everything." Or indeed that the symbol of moral pressure is the ever-watchful eye in heaven. For millennia, this link between behavior and being observed has been used by religions to make traditional societies more honest and fair. They remind us that our actions have consequences.

Just the thought that we are being observed is very persuasive. One can even think of conscience, our inner sense of right and wrong, as a gauge of how we will be viewed by others. Even two eyespots on a computer screen background are enough to boost generosity. Indeed, the electrical activity recorded emanating from the scalp of normal subjects has been shown to register more activity in response to isolated eyes than it does to full faces.

The effect was neatly illustrated by a little experiment carried out at Newcastle University, Newcastle upon Tyne in the UK. The common room in the university's psychology department had an "honesty box" in which fifty students, staff, and academics were asked to pay for tea, coffee, and milk. The system had been operating for many years, so users had no reason to suspect they were being used as guinea pigs in an experiment. Over ten weeks, the researchers placed a sign on the door of the cupboard where the honesty box sat above the kettle and coffeemaker.

Pictures of flowers alternated on a weekly basis with pictures of eyes—male or female, always looking directly at the observer. The expressions ranged from alert and watchful to manic. Every week the money collected in the honesty box was counted up. On weeks when the eyes image was shown, takings were almost three times more than during the flower weeks. The eye pictures were probably influential because they made the coffee drinkers fret about what others would think of them. There's evidence that a robot with large, humanlike eyes can have the same effect. The eyes seem to make us more aware that if we advertise we are good, we improve our chances of being helped at some future date.

Manfred Milinski and the economist Bettina Rockenbach described the remarkable nested effects of gaze on the watcher and the watched: "Observer Alice should take into account that the behavior of Bob (the observed) changes and therefore should conceal her watching; Bob should be very alert to faint signals of being watched by Alice, but he should avoid any sign of having recognized Alice's watching when switching from selfish to altruistic behavior. He should avoid turning his gaze in the direction of the recognized observer. On the other hand, as soon as Alice sees that Bob has recognized that he is being observed, she should eventually not reward the observed altruistic behavior."

Examples of this observer effect can also be found in nature. Take, for example, the cleaner fish we encountered in chapter 1. The cleaner wrasse gets its dinner by plucking parasites off the bodies of its "client" fishes, even from inside the mouths. The fish grooms its clients in the friendliest way when other client fish watch, but without an audience it is sorely tempted to nibble off pieces of its client's skin. In a similar way, experiments reveal that in a so-called Dictator game, where a person has to give away money to another, the amount they share drops by 50 percent if the recipient is unable to identify the donor.

When people behave in a charitable way, it reveals much about the fact that their behavior has been honed down the generations by ancestors wanting to make a good impression whenever they find themselves in circumstances where they suspect they are being watched. This need to impress was felt as keenly in a close-knit hunter-gatherer clan as in today's surveillance society. As we are about to see, the knowledge that our behavior is being observed—or that it may be observed—could provide policy makers with new leverage to deal with climate change.

HARNESSING REPUTATION

Words have a longer life than deeds.
                —Pindar

A simple message has already emerged from my research on the Tragedy of the Commons. Whenever individual behavior is relevant to the public good, it should itself be made public to help avert tragedy. Advertising is critical. When playing a public goods game, others have to know that you are doing your bit for the world. Only then can an individual's regard for his or her own reputation be fully exploited.

With my colleague Thomas Pfeiffer I tried to flesh out some examples of what this would mean for the ultimate Tragedy of the Commons, climate change. We need new ways to advertise how people behave. Domestic appliances already carry energy ratings. This idea should be extended as broadly as possible. Energy costs of individual households could, for example, be published by local newspapers. Companies could be ranked according to their emissions and their investments in climate protection. In America, where gas guzzlers persisted long after Europe and Japan, where new technology allowed them to be replaced by more efficient engines, stickers could be used to mark out the polluting vehicles with pitiful efficiency.

The bottom line of our experience with automobiles is that it is not enough to develop clean technology, we have to encourage people to use it too, just as Hardin realized long ago. Certain cars could have mandatory stickers, similar to those used on cigarettes, such as WARN- ING: THIS CAR IS HIGHLY INEFFICIENT; ITS EMISSIONS CONTRIBUTE TO LUNG CANCER AND HAZARDOUS CLIMATE CHANGE. Exposing who on your block or in your office uses the most energy might be a good incentive for everyone to reduce their carbon footprint.

Although these types of policies could raise issues related to privacy rights, the potential gains for the environment could be great. In the summer of 2006 there was an extreme drought in my hometown and everyone was repeatedly asked to cut water consumption. It was no longer legal to water our gardens. But then it emerged that thirty of the one thousand houses in my town consumed a significant fraction of all the water in the neighborhood. The local newspaper ran an article with the headline exposing the "thirsty thirty." The article read: "We know that five of the top 30 (and two of the top three) reside on Stratford Way. Two live on Weston Road, two on Sandy Pond Road, two on Tower Road and none live in North Lincoln. At least 6 of the top 10 have in-ground pools and one also has a whirlpool. Another has a hot tub but, alas, no pool. Most of the top 10 have either five or six full bathrooms plus at least two half bathrooms."

Many of the citizens of my town could figure out who the water hogs were. And if the water hogs realized this, I am sure that they made plans to cut their consumption accordingly. It struck me as an interesting example of how to make people cooperate. Knowing who uses what resources will allow those who contribute to reap reputational benefits, helping to compensate them for the costs they incur. When people publicly display their commitment to conservation, it is likely to increase the social pressure on free riders to do the right thing. A realignment of the internal compass of millions of individual minds can do much to augment government policies.

Many organizations are already becoming wise to this way of thinking. Hybrid cars such as the Toyota Prius have easily recognizable designs, which in effect advertise their driver's commitment to cleaner energy. Volunteers to environmental cleanup days receive T-shirts advertising their participation. In a scheme run by a local electricity company that was adopted by my colleague David Rand, if you chose to pay more to tap into electricity generated by alternative means, such as wind, you were given a "gone green" flag to plant in your garden.

Like it or not, billions of us are involved in the very real game of global warming. Even if we avert dangerous shifts in the global climate, we are still likely to face more extremes in climate and weather in the short term. Droughts, torrential downpours, heat waves, and floods are likely to occur more frequently. Sea levels will rise, along with the risk of extreme storm surges. Much more can and should be done to harness the power of reputation to encourage us to cooperate to avert dangerous climate change. This is one public goods game that none of us can afford to lose.

CHAPTER 11

Punish and Perish

If people are good only because they fear punishment, and hope for reward, then we are a sorry lot indeed.

—Albert Einstein

Down the ages, society has adopted one apparently simple and straightforward tactic to make people cooperate with one another: threaten them with punishment. Do as I say, or else you'll get it, whether a stiff rebuke, an eye-watering fine, imprisonment, beating, caning or flogging, or worse. That is why, to set an example for us all to follow, evil is usually punished in the happy—and somewhat gory—endings of fables, fairy tales, and myths.

Greek mythology had angry deities of vengeance and retribution, known as the Erinyes. Although these fearful goddesses were likely born as personifications of curses placed on the guilty, legend has it that they emerged from drops of blood that was shed when the mighty Titan Cronus used a jagged sickle to castrate Uranus, his father. The Romans called these doom-laden goddesses of revenge the Furies. In their quest for justice they appear in various guises. Their heads were wreathed with writhing black serpents, their eyes dripped with toxic blood, and their breath was burning hot. In the hunt for the guilty, their lust for punishment knew no bounds.

Fairy tales also brim with stories of revenge, at least before Disneyfication. Think of the death dance of Snow White's wicked stepmother,

221

the queen, when she is forced to wear burning iron shoes; when the hunter slices open the cunning wolf in "Little Red Riding Hood"; and the cannibalistic witch in "Hansel and Gretel" who was toasted and roasted in an oven. Could it be that we need an ever-present threat for us all to get along? Is this the way to avert the Tragedy of the Commons? Could it be that we cooperate better when under duress?

There's certainly plenty of evidence in popular culture of a link between punishment and reciprocity. Tit for tat, the instinct to answer one hurt with another, propels ancient Greek and Renaissance English tragedies, as well as eighteenth-century opera and nineteenth-century novels, toward blood-drenched denouements. When it comes to exploiting the dramatic potential of punishment, Hollywood is no different.

We have a repetitive and unquenchable appetite for spectacles of vengeance. *The Punisher* is just that: a former black-ops specialist for the government crusades against the bad guys who wiped out his family. Charles Bronson and Clint Eastwood have solved many problems with fists and bullets, only to ask questions later. There's rafts of revenge to be found in Roman antiquity (*Gladiator*), in Depression-era Chicago (*Road to Perdition*), and in postapocalyptic Australia with *Mad Max*. The story of "getting even" is ubiquitous. Just watch a soap opera or surf the web and you will find plenty of lurid examples.

Brain-scan studies have revealed what goes through the head of someone as they mete out a punishment. A liberal lashing with a stick is actually a carrot too: punishers seem to enjoy a warm, satisfying surge of blood to the reward centers of the brain. The flickers of activity observed in vengeful brains reveal there's satisfaction in comeuppance, a heady and intoxicating pleasure in exacting revenge, a neural correlate of schadenfreude.

The idea of punishment is always at the back of our minds because in the Darwinian struggle for existence, the winners live and the losers are punished with extinction. How does punishment fit into the story of cooperation?

In this chapter, I want to focus on retribution and payback and all those one-on-one interactions where one person punishes another for his misdeeds. This includes people who take the law into their own hands, or those who pay one person to have another killed. This "peer punishment" is thought by some to be an effective method for promoting cooperation. But this view is mistaken in my opinion. Peer punishment is problematic, and there are more effective ways to make people cooperate. I also want to make clear that punishment is not, as some have claimed, a mechanism for the evolution of cooperation. Instead, it fits neatly into the framework of reciprocity, which I investigated in chapters 2 and 3.

We have, for example, already talked about punishment in the context of Tit for Tat, when one defection is met with another defection. Then there is the strategy of Generous Tit for Tat, which also punishes defection with defection, though it can sometimes forgive and forego the punishment. As we saw in chapter 1, an even harsher strategy is called Grim: I cooperate as long as you do. But put one foot wrong, and I will defect forever. Punishment fits neatly into the framework of the standard Prisoner's Dilemma.

The payoff matrix in the Dilemma is such that "defection" could mean one of three things: First, withholding reward. So instead of cooperating, I do nothing at all. Think of the student who refuses to cook dinner for his flatmates when one of them refuses to do the washing up. Second is stealing. Now I take something from you and, as a result, you suffer a loss and I have a gain. Third, there is costly punishment, where I suffer a loss but, as a result of this, you suffer a bigger one. Perhaps a neighbor opposes your planning application because you did not keep your dog quiet at night. The neighbor is prepared to suffer a cost—your opprobrium and his time—to inflict a much bigger one in terms of denying you your dream of building an extension on your house.

Another possibility for incorporating punishment is to extend the Prisoner's Dilemma from two possible moves—cooperation and defection—to three: cooperation, defection, and punishment. Using a simplified form of the Dilemma, we can define them as follows. "Cooperation" means paying a cost for the other person to receive a benefit. "Defection" means doing nothing. "Costly punishment" means paying a cost for the other person to incur a cost. In punishing experiments, a ratio of 3:1 is typical, so that I pay one dollar for you to lose three. It is not hard to think of a real-world example of when we have to pay a cost to exert a punishment, such as the exertion it takes to cut down a neighbor's hedge to punish him for blocking the light.

COSTLY PUNISHMENT

The idea that costly punishment might be a powerful force to promote cooperation was suggested by an experiment done by two Austrian economists: Ernst Fehr, who is at the University of Zürich in Switzerland; and Simon Gächter, who was then at the University of St. Gallen and is now at the University of Nottingham in the UK.

Fehr and Gächter enrolled 240 undergraduate students to participate in the experiment. First they carried out a public goods game, of the kind discussed earlier. Then the players had the option to punish each other. They were all shown what everyone else in the group had contributed and then asked if they would pay to make another player lose even more, if that player had been stingy. The identities of the players were reshuffled each round. As a result it was not clear who was playing who from round to round, so the game was not truly repeated and there was no way that a player could develop a reputation. Moreover, you never found out who punished you. Hence there was no way to retaliate.

In all, they carried out ten experimental sessions with twenty-four subjects. Each subject played two six-period public goods games: a game without a punishment opportunity and a game with a punishment opportunity. When there was no punishment, players started out by being generous but quickly learned to be stingy. They found that they can punish free riders only by reducing their own contributions. Thus cooperation quickly unraveled.

What happened when punishment was introduced? Of the 240 participants in the experiment, 84 percent punished at least once and around 9 percent punished more than ten times. Most (74 percent) acts of punishment were imposed on defectors (that is, below-average contributors) and were executed by cooperators (that is, above-average contributors). With the help of punishment, cooperation among the players held good: more than 90 percent of the participants contributed more money when punishment was a possibility. Because this form of punishment saw the fines return to the experimenter, not to the punisher, it was costly. Yet Fehr and Gächter showed that, nevertheless, the tendency to engage in this form of punishment was widespread. Even in the last round of the game, when there was actually no point in punishing to encourage future cooperation, punishment still took place.

Based on this, the researchers proposed that people "like to punish" and that costly punishment was a new mechanism to promote cooperation. They even coined the term "altruistic punishment" to indicate that our urge to punish is present even when we ourselves do not benefit from the action. The term suggests that we are willing to punish others for the greater good and the benefit of society at large.

While I find the Fehr-Gächter experiment fascinating and important, I disagree with some aspects of their interpretation. First of all, costly punishment is not a separate mechanism for the evolution of cooperation. If I punish you for defecting during an earlier encounter with me, then it is direct reciprocity. If I punish you for defecting in games with other players, it is an example of indirect reciprocity. Second, the motivation of people who mete out costly punishment in real life is hardly ever "altruistic." Harming or hurting others marks an escalation of conflict. This punishment is motivated by anger, greed, and aggression. It is primarily used to hold others down, to exploit them, to weaken them, to get rid of them. The term "costly punishment," which indicates that the punisher incurs some expense too, seems more appropriate than "altruistic punishment."

A close look at the results of these experiments revealed that the average payoff in the punishment group was lower than in the control. Punishment could indeed force cooperation in the public goods game but at a cost that was so high that it destroyed the advantages of cooperation. In other words, it would have been better not to offer the possibility of punishment at all.

The game also had a somewhat artificial design. One round of the public goods game was followed by one round of punishment. Players were informed who had contributed how much in the public goods game, but players were not informed who punished them. Therefore punishment could be dealt out anonymously and with no fear of retaliation. The experiment was designed to make punishment work as effectively as possible. Real life is, of course, very different. When we retaliate against our peers, they usually know who is punishing them and we must steel ourselves for reprisals.

When in our evolutionary history had our ancestors encountered the peculiar situation seen in the Fehr-Gächter game, in which we know exactly who contributed what to the public goods game, but not who punished us in the subsequent round?

Evolutionary psychologist John Tooby of the University of California, Santa Barbara, points out that for the vast majority of the period during which the human mind has evolved, people lived in small groups and encountered others repeatedly. In ancient societies, you would know which rival tribesmen did what to whom. As a result, in order to stay on good terms with others whose help they might need someday, there was pressure to cooperate. Even in an anonymous modern setting where a one-off encounter is likely—say on a highway or in a city—most of us have the foresight to see that the stranger who bumped into us or the cyclist who almost ran us down could turn up again, in the workplace, at a party, or wherever.

So-called one-shot experiments, which try to exclude the factors of repetition and reputation, are difficult to interpret. The reason is that they are unrealistic. All our instincts, intuitions, and behaviors have been shaped over the generations by situations where we encounter the same people again and again, and where reputation does play a role. Thus it is dangerous to infer universal truths about human behavior from one-shot experiments. Robert Trivers, in discussing the Fehr-Gächter experiment, commented that biologists take a spider to the lab to study what makes it tick but that does not mean that biologists think that the spider has evolved for life in the lab.

If punishment is typically used in the context of repeated dealings where people know who has punished whom, then it is difficult to draw conclusions from an experiment where punishment occurs anonymously. If we want to understand human behavior and the interplay between cooperation and punishment, we need to study situations where the players encounter each other again and again. It makes little sense to study human behavior outside the sturdy frameworks of direct and indirect reciprocity. All the crucial interactions of everyday life occur in the context of repetition and reputation.

WINNERS DON'T PUNISH

This was the Golden Age that, without coercion, without laws, spontaneously nurtured the good and the true. There was no fear or punishment: there were no threatening words to be read, fixed in bronze, no crowd of suppliants fearing the judge's face: they lived safely without protection.
—Ovid, *The Metamorphoses*

One day I was visited at Harvard by a prospective graduate student from Sweden who was keen to do laboratory experiments. Anna Dreber was born and raised in the city hall of Stockholm. Her mother, a formidable feminist, politician, and ethical activist, had been the vice-mayor of the city at the time. Anna had spent her formative years in the town hall, which had been inspired by the palaces of the Renaissance and was built in the 1920s around two squares, or "piazzas," Borgargården and the Blue Hall. Each December as the dark Swedish night was brightly lit by chandeliers, Anna would witness all the glitz and hoopla that surrounded the Nobel banquets.

Now she sat before me, suggesting that she could do a doctorate with me to add some real-life experiments to our portfolio of evolutionary game theory based on theory and computation. Within a few weeks, Anna had discovered that we were able to apply to use the Harvard Business School's Computer Lab for Experimental Research. A short walk away, the lab turned out to be a modern facility full of screens and smart blue booths, where students could be lured by small sums of money to play games of any shape, size, or format to explore cooperation.

We decided to stage a more balanced experiment to evaluate the impact of punishment. We both felt that most experiments were designed to outline the positive effects of punishment. Now we wanted to do experiments where the negative effects were not suppressed. This, of course, depended on the approval from an ethical committee. We were amused to find that we could not do any experiments using deception in the business school: they could only be done in the psychology department. Fortunately, the business school did allow us to use punishment.

At one party in Nowakia—festivities were commonplace before the global financial crisis—I introduced Anna to Dave Rand. They got on famously and decided to work together. Dave came from a quite different background. He grew up around planes. His father, a professor of applied mathematics at Cornell University, is a hobby pilot and allowed his son to steer from a very young age. To be a good pilot you need a clear mind, and this might explain why Dave possesses an uncanny amount of common sense. He was the first Harvard graduate in systems biology, both a theoretician and an experimentalist, who now works with psychologists, economists, and law professors. His talents do not end there. He is a one-man rock band. He plays a mean guitar. You can find his recordings on YouTube.

Because extraordinary circumstances can generate extraordinary behavior, David, Anna, and I decided to see if repeating encounters between players made a difference to the effectiveness of costly punishment. Working with Drew Fudenberg of Harvard, a leading expert in economic game theory, we asked 104 college students to take part in a variant of the repeated two-player Prisoner's Dilemma. We made pairs of subjects play repeated games with three options. The two players had a choice between cooperation, defection, and costly punishment. "Cooperation" meant paying one dollar for the other person to receive three dollars. "Defection" meant taking away one dollar from the other person. "Punishment" meant paying one dollar for the other person to lose four. The important difference between our setup and earlier experiments was this: we allowed them to slake their thirst for revenge. If Alice punished Bob, then Bob had the opportunity to punish Alice in the next round.

We monitored 1,230 repeated interactions between pairs of players, each lasting between one and nine rounds. It was fascinating to observe. In the second figure of the *Nature* paper that carried the results, we showed the effects of costly punishment. What was gratifying was that nice people finished first. A pair who cooperated over four rounds were equally ranked first in terms of payoff. Those who turned the other cheek did well too. When one person cooperated in two consecutive rounds, despite being matched with defection each time, he or she went on to cooperate and ended up being sixth overall. The defector, meanwhile, was converted to cooperation in the last three rounds and ended up nineteenth. A cooperator was so enraged by a defection that he or she responded with punishment. After five rounds, the defector had not cooperated in the face of punishment. They ended up twenty-fifth and twenty-second respectively.

There was also the cooperator who when confronted with a defector responded with punishment. That triggered retaliatory punishment from the defector, then round after round of punishment and counterpunishment. The players of this particular game of mutually assured destruction were ranked thirtieth and twenty-fifth. In one case we had evidence of the consequences of preemptive strikes. After mutual cooperation, one person punished and this again triggered mutual defection. The punisher was ranked twenty-ninth and his fellow player twenty-fourth.

What leapt out at us from the study was the clear link between punishment and poor results. The six best-performing people never used punishment. In contrast, the worst-performing people used punishment most often. Thus winners do not punish, losers do. The study showed that players who earned the lowest payoffs punished most often to trigger a downward spiral of retaliation, with destructive outcomes for everybody involved. So Henry punishes a defection by Sally, which triggers counterpunishment. Then Henry is fed up and deals out more punishment. And so on.

Perhaps this outcome was a fluke. Perhaps our winners were simply lucky in the sense that they found themselves with a cooperative opponent. We did a deeper analysis, measuring the probability that a player would respond to defection with punishment. Those who did well did not escalate fights with a retaliatory punishment but just stuck to defection, meeting like with like. The message was clear. Punishers don't win. Thus, when repetition is taken into account, punishment by people who like to take the law into their own hands is ineffective.

REWARD IS BETTER THAN PUNISHMENT

Our first foray into this study had raised questions about the wisdom of punishment in the situation of repeated encounters between pairs of players. But the two-player game we studied was not a public goods game, which can be thought of as a Prisoner's Dilemma with *more* than two people. Therefore our results did not map easily onto previous studies, such as the one by Fehr and Gächter, most of which had concluded that punishment is more effective than reward in maintaining cooperation in public goods games. So we decided to stage our own public goods game in collaboration with Tore Ellingsen, a Norwegian who works at the Stockholm School of Economics and is one of Sweden's most celebrated economists.

In all, 192 people took part in this next experiment. They were divided into groups of 4, with 16 control groups and 32 experimental groups. In effect, we followed the original Fehr-Gächter recipe but with an extra ingredient. We allowed repeated rather than anonymous encounters, so there was an opportunity to know who was punishing whom.

We had three "treatments" and one control experiment. The control was a standard public goods game. Treatment one allowed punishment, treatment two allowed reward, and treatment three allowed punishment and reward. The costs were as follows: "Punishment" meant you paid something for someone to lose something. "Reward" meant you paid something for someone to gain something. Each group had a marathon session of fifty rounds.

All three treatment groups saw cooperation emerge, unlike the control group, where it floundered in a traditional Tragedy of the Commons. But even though punishment led to cooperation, it was costly and the total payoff was as low as in the control group. The reward treatments also maintained high contributions in the public goods game, but, significantly, the total payoff in the reward session was much higher than in the control session. We found that if both reward and punishment are on offer, then the winning groups do not use punishment, which turns out to be both costly and ineffective. Rewards go further than punishment in both benefiting the public good and in building cooperation, despite the efforts of free riders.

A simple idea would emerge from this experiment: the Tragedy of the Commons can be solved by linking the public goods game to games with targeted interactions. By this I mean that, rather than withdraw your cooperation, which affects all players in a traditional public goods game, you withdraw it only from those that defect and, even better, reward those that cooperate. Cooperators in the public goods game gain a reputation, which makes them more attractive partners for other cooperators in private—one to one—dealings, just as a company with good green credentials will find it easier to win business. This recipe for cooperation is simple and effective. When our paper appeared in the journal *Science* in 2009, it was entitled "Positive Interactions Promote Public Cooperation."

ANTISOCIAL PUNISHMENT

Another experiment that cast doubt on the efficiency of costly punishment in promoting cooperation was performed by Benedikt Herrmann, Christian Thöni, and Simon Gächter. They studied the behavior of people in sixteen cities, from Boston and Bonn to Riyadh, Minsk, Nottingham, Seoul, and others, marking at that time the largest cross-cultural study of experimental games in the developed world.

As before, they played a public goods game in which participants were given chips and told they could either keep them all for themselves or put them into a common "pot" that would yield extra interest that would be shared out equally among all players. Over ten rounds of the game, 1,120 middle-class college students in Boston and Copenhagen contributed about eighteen chips each; those in Athens, Riyadh, and Istanbul, only six. The most cooperative participants, who contributed up to 90 percent of their chips, contributed 3.1 times as much as the least cooperative, who had an average contribution of 29 percent of their chips.

Behavior varied dramatically when the players were given the ability to punish another by taking tokens away. As the earlier work had shown, players were willing to part with a token of their own in order to punish the low investors or the freeloaders who had exploited others. But in the international version of this game, striking national differences would arise.

When freeloaders were punished for putting their own interests ahead of the common good in countries such as the United States, Switzerland, and the United Kingdom, the freeloaders accepted their punishment and became more cooperative and the earnings in the game increased over time. However, in countries such as Greece and Russia, the freeloaders sought retribution. Because they used the same setup as the original Fehr-Gächter experiment, there was no possibility for targeted revenge. Instead they lashed out generally at cooperators. Presumably they reasoned that they could preempt their next punishment this way, or reasoned that they were the likeliest people to have punished them in a previous round. Perhaps free riders punished cooperators as a show of dominance, to say something like "These cooperators are fools, stupid and weak for not keeping everything for themselves, and I will punish them to show them who is boss."

The study seemed to confirm the stereotypes that the British have a sense of fair play, while the Greeks thirst for revenge. Players in Athens and Muscat had the highest level of revenge punishments, retaliating against the enforcers—punishing cooperators—about six times as often as did students in Seoul, Bonn, Nottingham, and other cities. Samarra, Minsk, Istanbul, and Riyadh were somewhere in the middle.

What is also fascinating is that the thirst for retaliation and dominance seemed to track measures of the norms of civic cooperation and rule of law that had been made by social scientists in what is called the World Democracy Audit. These norms covered general attitudes toward the law—for example, whether or not citizens think it is acceptable to dodge taxes or flout rules—and weighed up political rights, civil liberties, press freedom, and corruption. In societies where public cooperation is ingrained and people trust the police and their law enforcement institutions, revenge is generally shunned.

But in those societies where the rule of law is perceived to be ineffective—that is, if criminal acts frequently go unpunished—antisocial "revenge" punishment thrives where a defector punishes a cooperator. Cooperation is very strongly inhibited as a result, so there is an incentive to take a free ride and ignore civic-minded initiatives such as recycling, neighborhood watches, voting, maintaining the local environment, tackling climate change, and so on.

Importantly, the work also revealed that punishment did not always increase cooperation in subsequent rounds of the international games. Cooperation in about half of the participant pools remained at the initial level, and the higher the level of antisocial punishment in a participant pool, the lower was the rate of increase in cooperation. At best, "altruistic punishment" did not help people to cooperate very much. This seemed to me to capture something of the flavor of real life.

THE POINT OF PUNISHMENT

Let's stand back and set this work on punishment in a broader context. There are two basic kinds of punishment. In this chapter, I have focused on one of them: peer punishment, the kind used by the Mafia, or in instances when people take the law into their own hands. The second kind of punishment comes from a higher level of authority, typical of that found in a hierarchy; for example, it could be the state punishing people who break the law, whether by hanging, imprisonment, or lethal injection; it could be parents punishing their children in the name of education; it could be a boss punishing her workers; or it could be a means to maintain hierarchy, so that a captain can put a foot soldier in his place, and so on and so forth. This is "institutional punishment"; I am also interested in this form of punishment and it is important. But there have been few game theory studies to shed light on what is going on.

So far most of the focus has been on peer punishment. We need to see it in the context of the Tragedy of the Commons and the Prisoner's Dilemma. In a two-person Prisoner's Dilemma, I am able to punish a defector by switching from cooperation to defection. But when lots of people play, as they always do on those tragic commons, by switching to defection I harm both the cooperators and the defectors. I harm everyone. Therein lies the problem.

There's no way to target people in standard public goods games. You have the chance to give or not to give to a public pool. But other players have no opportunity to interact directly with you—whether to punish you or reward you or whatever—because the experimenter takes the pot, multiplies it by some amount, and then redistributes it among the players. If you are irritated by how little one person puts into the pot, your only sanction in a classic public goods game is to cut the amount you put in and, as a result, punish every other player.

If players have the ability to deal directly with each other, then the problem of the public goods game can be solved. We have seen that punishment is one mechanism, though it's very inefficient. The prize of costly punishment often outweighs the increased cooperation in the public goods game (even in the experiments that were designed to suppress retaliation against punishers). A much better way to solve The Tragedy of the Commons in these games is by positive interactions between players—that is, to reward fellow cooperators by establishing mutually beneficial private interactions with them. Then the public cooperators gain a reputation that makes them more attractive prospects to fellow cooperators. In this way, private cooperation can bootstrap public cooperation.

There is another interesting corollary to the idea that a targeted response is more effective. Let's say that you want a workforce of two hundred engineers to build a fast, luxurious automobile. Each of them plays a role in the production process and so, to make them more efficient, you decide to introduce a regime where the failure to add a component, say a screw, leads to a fine. I am sure that this form of punishment would make the engineers take more care when installing the components. However, I am also sure they will do the minimum necessary to fulfill their contract.

But what if you reward them for success—for instance when you sell more automobiles? If you allow the engineers to take a share of the profits, you may well find that they are inspired to do much more than ensure the correct component is added at the right time. Instead of simply meeting their obligations, they might come up with new production processes and regimes. They might reorganize the flow of components, or find ways to fit more than one at a time. Reward leads to more creative forms of cooperation than punishment. Reward does more than make us work more effectively together—it stimulates creativity too. Reward, not necessity, is the true mother of invention.

CHAPTER 12

How Many Friends
Are Too Many?

*The world is so empty if one thinks only of mountains, rivers and cities;
but to know someone here and there who thinks and feels with us, and
though distant, is close to us in spirit—this makes the earth feel like an
inhabited garden.*

—Johann Wolfgang von Goethe
Wilhelm Meister's Apprenticeship

Here's a fascinating—and humbling—social experiment. Select one hundred people at random in New York City and ask them each to list all their friends, so you can figure out their average number of friends. Then, in turn, ask their friends how many friends they have. You will find that the latter's average number of friends is higher. Sociologist Scott Feld of Purdue University, West Lafayette, drew attention to this apparent paradox in a paper that he baldly entitled "Why Your Friends Have More Friends Than You Do." The explanation lies in the realization that there is a bias in the question being asked, since you are more likely to know popular people and less likely to know unpopular ones. That is also why your lover has had more lovers than you. Or indeed why people at your local gym tend to be fitter than you, because you do not encounter the relatively out-of-shape ones who rarely turn up.

There is a moral in this story of apparent social inadequacy. If we want to understand the role of cooperation in evolution, we have to understand how it is bound and guided by the structure of populations.

The word "evolution" comes with important baggage that is often neglected in popular discussions. Evolution is about change in a population, not just in a solitary creature. When some reproducing individuals in the population become fitter, they are more likely to survive and produce offspring. Over successive generations, these fitter individuals become more common. In this way, "evolution" refers to a change in the genetic composition of a population. A corollary of this is that the structure of a population can deflect the trajectory of evolution.

The word "structure" may sound a little abstract when applied to a population, but it can have important effects. To illustrate the way I think about population structure, I want to draw on terms that are common in classroom chemistry and physics, notably in discussions of what happens to states of matter as we raise the temperature.

Let's start by being relatively chilly. There are the serried ranks of molecules or atoms in a solid, where the relationships in the structure are fixed. The warmer liquid state sees fleeting and fluid associations among molecules. These associations have evaporated by the time we reach a gas, where component molecules speed here and there, hither and thither.

Let's keep these ideas about states of matter at the back of our minds to describe the structure of a population. The one that is most familiar is a population where the individuals are people. But, of course, they could also be birds, cells, bacteria, and molecules—indeed, anything and everything that is capable of interacting with something else in some kind of community, whether a flock, body, biofilm, or primordial pizza. Now let's explore how to think about the various states of population.

A "well-mixed" population is featured in earlier chapters. You can think of this as a gas. Individual players bump and knock into each other at random, just as individual molecules of gas bash into each other in a disorderly way. More precisely, mathematicians like to say that in such a population every player is equally likely to interact with every other. We saw in earlier chapters that unless players are smart (in which case they can resort to direct and indirect reciprocity) cooperators always lose out to defectors in well-mixed populations.

There is a population structure that, like a liquid, is between the two extremes of gas and solid. I'll deal with this important state of biological matter in the next chapter, when I discuss games on sets. At the other end of the extreme is the equivalent of a solid. We saw one way to describe such populations, where the relationships among players are fixed, when we encountered spatial games in chapter 3. In these games, the players are connected by geography. Those that interact happen to be next to each other. In a society of hunter-gatherers, for example, the range of the players could be limited by roads, rivers, and mountains. People tend to deal with adjoining villages, settlements, and so on. Now cooperators can cluster together and thereby defend themselves against exploitation by defectors. In these populations, strategies of cooperation and defection can coexist in uneasy standoffs or, more often, in cycles of boom and bust.

But, of course, in modern society we can interact with players all over the world, courtesy of the telephone, email, or internet. These networks are messy and ubiquitous, from the myriad links between multinational businesses to the dizzying interconnections that straddle the planet in the form of the World Wide Web. We call these complicated arrangements networks but the general term that mathematicians use for such a structure—whether social, transport, neural, or whatever—is a "graph." The individuals in the graph are called the vertices, or nodes. If two individuals know each other, then they are connected on the graph by what is called an edge. In this chapter, we will deal with games on graphs.

We are today embedded in a vast, complex, sprawling network of family, friends, colleagues, business partners, and so on. These graphs swaddle the world. I want to show you just how connected our world is and why this is important. Then I will erect a theoretical framework to generalize the mechanism we encountered before, in the earlier chapter on spatial games, to reveal how cooperation can thrive when we play games on graphs, such as social networks, and in much broader classes of population structures.

BRIEF HISTORY OF NETWORKS

Fascination with social networks dates back decades. The evocative suggestion that we are all linked to each other through a chain of mutual acquaintances is often credited to the Hungarian author and playwright Frigyes Karinthy, who mentioned the idea in his short story "Chains," published in 1929. But when it comes to the scientific study of these networks, perhaps the best-known piece of research is one of the earlier investigations, carried out by American social psychologist Stanley Milgram at Harvard in the 1960s. In one experiment, Milgram sent packages to 160 people randomly selected in Omaha, Nebraska, asking them to forward the package to a friend or acquaintance who they thought would help bring the package closer to a target person, a stockbroker who lived in Boston, Massachusetts. Amazingly, given the tens of millions of people in America, his experiment suggested that there tended to be just six people on average linking one person with any another—giving rise to the popular notion that we all may be connected by just six degrees of separation.

Milgram's original work was criticized for various shortcomings. Still, it proved to be highly influential, not only in science but in culture too. The idea features in the John Guare play of the same name and the 1993 film version featuring the Hollywood actor Will Smith. In 2006 there was the TV show *Six Degrees,* which told the story of six characters who, according to the network, "go about their lives without realizing the impact they are having on one another." My coauthor Roger Highfield has carried out a couple of pop experiments along the same lines with Richard Wiseman of the University of Hertfordshire, as a way to highlight that we do indeed live in a small world.

Running in parallel with this have been attempts to quantify the degree to which we are connected. One mathematician in particular has come to symbolize this work: the extraordinary, itinerant Paul Erdős. Born in 1913 in Budapest, Erdős was driven out of his mother country as a young man by anti-Semitism. He moved to Manchester, England, in 1934, the same year that he was awarded a doctorate in mathematics. So began his travels back and forth to work with academics at institutions far and wide until his death, fueled all the time by a potent blend of curiosity, coffee, and amphetamines.

Erdős worked on a wide range of topics, generating around 1,500 articles in his remarkably prolific lifetime. Among his five hundred or so coauthors was Alfred Renyi, the source of the excellent and truthful quotation "A mathematician is a device for turning coffee into theorems." In 1959, they modeled the networks seen in communications and the life sciences by connecting their nodes with randomly placed links. This work is perhaps the most relevant to this chapter, but when it comes to networks Erdős is also known for the idea of the Erdős number—which measures the "collaborative distance" in authoring papers.

The lower the number, the closer a person is to Erdős. This is a matter of considerable pride among mathematicians. We start with his own number, which is 0. Erdős's coauthors have an Erdős number of 1. People other than Erdős who have written a joint paper with someone with an Erdős number of 1 but not with Erdős himself have an Erdős number of 2. By the same reasoning, my own Erdős number is 3. If you count books, Roger Highfield's is 4. And so on and so forth. If there is no chain of coauthorships connecting someone with the great Erdős, then that person's Erdős number is said to be infinite. A similar exercise can be conducted for any other individual. One well-known example is the game Six Degrees of Kevin Bacon. And there's even the combined Erdős-Bacon number for those who want to bridge the apparently unconnected worlds of mathematics and acting.

In the wake of such efforts to show the degree to which people interact came an interesting paper that put Milgram's observations into a theoretical framework assuming that we live in a small world. Duncan Watts of Columbia University and Steven Strogatz of Cornell University proposed a mathematical model of a network in which each point, or node, is closely connected to others nearby and, in addition, there are some long distance connections. The six degrees of separation idea works in such a network because in every small group of friends there are a few people who have much wider connections, either across continents or across social divisions. Small worlds can be found in situations as diverse as networks of people, power grids, the World Wide Web, and the neurons in the brain of a nematode worm.

A fascinating property of such networks is that they often seem to be scale-free, as emphasized by another Hungarian, Albert-László Barabási, who is now the director of the Center for Network Science at Northeastern University. You can build these networks by adding nodes, one at a time. The new nodes have a higher probability of linking to nodes that already have many connections, thus the "the rich get richer." This means that the distribution of links approximates to what scientists call a "power law," where a tiny fraction of nodes receive a hugely disproportionate share of links, and the vast majority are mostly ignored. These networks have a small number of hubs that are significantly more connected than the other nodes in the network. Since the number of links to a website closely relates to its popularity, traffic, and search-engine ranking, these studies imply the "winners" will continue to dominate the web (for example, sites like Google and Amazon), while new entrants find it hard to compete. This is the well-known "winners take all" concept.

There are practical implications of this understanding. Viruses can spread efficiently across such networks, as illustrated by the work by Roy Anderson and Bob May on the spread of epidemics in heterogeneous populations conducted in the late 1970s. Moreover, these networks are robust when it comes to random attacks, which are most likely to take out unimportant nodes because, of course, the vast majority of them are. By the same token, major disruption can be caused by well-informed cyberterrorists attacking the most highly connected nodes. Knocking out just a few of these hubs could halt the flow of information and rapidly fragment the web.

The small world question is still a popular topic today, with many experiments being conducted with email and social networking sites such as Facebook. There's popular interest in the idea too. One can even find applications that are able to work out the number of steps between any two members of a site. But before I turn to the subject of how networks affect cooperation, let's take a short detour to look at why this subject touches all of us.

ONLY CONNECT

A man should keep his friendships in constant repair.
—Samuel Johnson

We all know that we are affected by our networks of friends and relatives, in the sense that they can give us all kinds of things, from houses and birthday presents to colds. But what is fascinating is that there's evidence that they can pass on their state of mind too. Research by Nicholas Christakis of Harvard Medical School and James Fowler of the University of California, San Diego, suggests that we are swayed by the moods of friends of friends, and of friends of friends of friends—people several degrees of separation away from us whom we have never met, but whose disposition and behaviors can ripple outward toward us through an intervening social network.

Christakis and Fowler found that happy people tend to be clustered together, not because they gravitate toward smiling people, but because of the way that happiness spreads through social contact over time, regardless of people's choice of friends. When an individual becomes happy, a friend living within a mile experiences a 25 percent increased chance of becoming happy too. And for next door neighbors, that cheering probability rises to 34 percent.

Another surprise came with indirect relationships. Again, while an individual becoming happy increases his friend's chances of smiling, a friend of that friend experiences a nearly 10 percent chance of increased happiness, and a friend of that friend has around a 6 percent increased chance—a three-degree cascade of good humor. Thus, your actions and moods—whether blue or jolly—affect your friends, your friends' friends, and your friends' friends' friends. Fowler and Christakis have even done experiments in which strangers are randomly assigned to interact with each other, and they have found that altruistic, cooperative behavior also spreads to three degrees.

Thereafter your influence fades away from the network like the smile of the Cheshire Cat. "While all people are on average six degrees separated from each other, our ability to influence others appears to stretch to only three degrees," says Christakis. "It's the difference between the structure and function of social networks." Alison Hill and Dave Rand have studied the data to determine if happiness and depression behave like infectious diseases. It turns out that they do. They also found an interesting twist: happiness is more infectious than depression, so that the average lifetime of a happiness "infection" is about a decade, compared with five years in the case of unhappiness.

Christakis and Fowler have also shown how understanding social networks can help us chart real-world disease outbreaks. They followed up the "friendship paradox" I described at the start of the chapter: statistically speaking, the average person is much more likely to know a popular person simply because she has so many more friends. And when they monitored the spread of both seasonal flu and H1N1, swine flu, through students and their friends at Harvard University, they found that the infection rate peaked two weeks earlier among the better connected.

The work suggests that by studying the friends of a randomly selected group of individuals, epidemiologists can isolate popular people who are more connected and are therefore more likely to catch viruses early. This could allow health authorities to spot outbreaks weeks in advance of current surveillance methods. More evidence, as if we really need it, that the very structure of our own social network can have a profound influence on our lives.

EVOLVING NETWORKS

When I first took an interest in networks, I wanted to come up with a way to reveal how they shaped the evolution of cooperation. Initially, my interest was driven by my efforts to understand how the architecture of tissue in our bodies can make cancer less likely to take hold. This work, described in chapter 7, prompted me to get interested in the general question, How does population structure affect evolutionary dynamics? I began to work on this problem with Erez Lieberman and Christoph Hauert, a biomath and computer wizard from Bern, Switzerland, who was then in my group but is now at the University of British Columbia in Vancouver. Our collaboration gave rise to a new field called evolutionary graph theory.

As the above discussion of scale-free and random networks illustrates, graphs come in many forms, shapes, and sizes. There are those where each person is connected to his neighbor on a regular grid. There are also graphs where each person is connected to every other player. And there are all the networks with structures in between, from ordered to disordered and all sorts of mixtures of the two. How could we work out the effect of network structure on cooperation?

Our work on evolutionary graph theory began by studying what is known as constant selection. In other words, we considered the following simple scenario: Take a resident population and introduce a single new mutant, a variant of the existing residents. This new mutant could have a selective advantage, which means it reproduces faster. Or it could have a selective disadvantage, which means it reproduces more slowly. Or it could have the same rate of reproduction, in which case it is called a "neutral" mutant. And we wanted to find the answer to a simple question: What is the probability that the lineage of descendants arising from the mutant takes over the population? This quantity is known as the fixation probability of a new mutant.

One can ask this question in all kinds of circumstances. The individuals could be cells. Some are normal (so-called wild type), while others are mutants that could be on the way to cancer. The same question could also be posed in a cultural setting: What is the chance that, once you have invented a fad, others take it up? Even though this question sounds different, the basic issue remains the same. What is the probability that a mutant fad—from pop song to TV format to fashion— can multiply to take over the entire population?

This probability can be easily calculated for neutral evolution, which means that the resident cells and the new mutant have the same fitness. Then every cell has the same fixation probability of taking over and being the ancestor of the whole population some time in the future. So for 10 cells, there is a 1 in 10 chance of taking over. For 100 cells, there's a 1 in 100 chance. And so on and so forth. The fixation probability of a neutral mutant is the inverse of the population size.

If a mutant has a selective advantage (or disadvantage), then there exists a mathematical formula to describe its fixation probability in a well-mixed population. But we wanted to understand how the structure of graphs affected the fixation probability. We found that many graphs behave exactly like the well-mixed population: in other words, they do not alter the fixation probability of new mutants. Note that the well-mixed population itself is described by what is called a "complete graph," where every individual is equally connected to every other individual.

But in doing this, we also found amplifiers and suppressors of selection. Networks that act like amplifiers can augment the chances of advantageous mutants. That boosts their ability to take over a population. By the same token, suppressors reduce the chances of advantageous mutants. Depending on how they guide natural selection, these graphs have different anatomies.

Amplifiers are often starlike structures. The World Wide Web is one potential example, where there are hubs of highly connected individuals. These hubs are hotspots of evolution. Another example of an amplifier network consists of a funnel, where one node is connected to three, then nine more, and so on until the structure wraps back to our first node. Or even a metafunnel, where many funnels sprout from a single node, or a superstar, which looks like a many-petaled daisy. Structures like the superstar and the metafunnel are supercharged amplifiers of selection. They virtually guarantee the fixation of any advantageous mutant. In such a population, a good idea can never be lost.

Suppressors, on the other hand, tend to be organized into hierarchies. Small upstream populations that feed into large downstream populations act as suppressors of selection. In such populations, there is a good chance that innovations are overlooked. This kind of network can be found in the organization of tissues, for example. As I described earlier, in my work on stem cells, crypts, and cancer, there are many tissues in the human body with a population structure that damps down selection. A single stem cell divides and produces differentiated cells, which differentiate further until they give rise to terminally differentiated cells, which eventually die. All cells are the offspring of the stem cell but only the stem cell makes more of its own kind. In this way, we have evolved a tissue design that is meant to fight off cancer for as long as possible (in the context of a normal human lifespan).

These amplifiers and suppressors might have some potential in the future design of "evolution machines." Many fields of science have already taken advantage of evolutionary ideas. There have been various attempts at Darwinian evolution inside a computer, in which organisms—scraps of computer code—fight for memory (space) and processor power (energy) within a cordoned-off "nature reserve" inside the machine. Computer scientists have played around with evolutionary software that can gradually evolve and mutate to carry out a task efficiently, or hone the performance of a wing, robot, or whatever, without the need for a designer to get involved. One of my postdocs, Feng Fu, from Peking University, spans the fields of evolutionary dynamics and evolutionary robotics. Given the trend in computing and robotics to learn from biological examples, I like to think that our selection amplifiers and suppressors could find all kinds of uses in the brave new world of evolutionary robotics and in biomimetic machines.

In the world of management, which is always questing to find the most efficient corporate structure, we can also ask, for example, which networks are well suited to ensure the spread of favorable concepts. Here again selection amplifiers, built around stars or funnels, will enhance the spread of brilliant ideas arising from any one individual to ensure that they efficiently trickle throughout an organization.

NETWORK GAMES

When we survey our lives and endeavours, we soon observe that almost the whole of our actions and desires are bound up with the existence of other human beings.

—Albert Einstein, *The World As I See It*

So far we have seen the effect of networks (graphs) when individuals have a constant fitness. Next we want to study how networks affect the outcome of evolutionary games. The basic idea is similar to spatial games: individuals interact with their neighbors on a network and accumulate a payoff. The greater the payoff, the more chance that this individual will reproduce or that others will imitate his strategy. That sounds simple. But calculating "games on graphs" turned out to be unusually difficult.

I invited Hisashi Ohtsuki, who was at the time a postdoctoral investigator in my group at Harvard, to join the project. My encounters with Hisashi had always followed the same pattern, one that was both remarkable and reassuring. On day one I would discuss a problem with him. On day two he would return and say: "Martin, I have preliminary results." Sometimes, these preliminary results would consist of many pages covered with dense calculations in small, neat handwriting. Even though they were done in pencil, I was always struck by the fact that there were never any corrections.

Thinking of a scene in the Mozart movie *Amadeus,* I would press him: "These are originals?" He always replied, "Yes, originals." I would then ask, "Are you sure about these results?" He would always give me the same modest reply: "No, no. Only preliminary results." Then on day three he would usually come back and say: "Martin, I have final results." That simply meant he had checked his calculations and, as usual, had found no mistakes.

But to solve games on graphs, Dr. Ohtsuki (as I like to call my brilliant friend) did not return on the next day. At last, he had encountered a real challenge. He was wrestling with a problem that he could not solve immediately. He needed to draw on a wide range of mathematical techniques for the problem of games on graphs. In all, it would take him several weeks. By the extremely high standards of Dr. Ohtsuki, this was unprecedented.

Meanwhile, Christoph Hauert tackled the problem head-on by using his superfast computer to simulate what happens on various kinds of graph, revealing an intriguing trend. Christoph and I pored over the digital utterances of his binary friend. We used the computer to chart the evolution of cooperation on a variety of structures. These included cycles, regular lattices, random graphs of the kind pioneered by the great Erdős, random regular graphs, and scale-free networks.

Let's take a cycle, for example. Here each individual in the graph has two neighbors. As for a regular lattice, think of a chessboard, which takes us back to the work I described in chapter 3. When it comes to creating a random graph, as Erdős originally did, take a number of individuals, then connect any two with a given fixed probability. A random regular graph, on the other hand, is also a structure that is generated at random, but where you have taken care to ensure that each individual has exactly the same number of neighbors (this is a bit artificial but it can simplify calculations). Finally, there's the scale-free network I described earlier, which tends to have a few highly connected individuals and many individuals that have only one or two connections.

When playing evolutionary games in structured populations, it is important to specify the update rule. This is the rule that determines how individuals change their strategies. Different update rules can generate very different evolutionary outcomes. In our experiments, we used the following rule. At random, an individual is chosen to "learn" from his neighbors. He looks at all his neighbors and then tries to imitate one of the strategies, with a likelihood that is proportional to payoff. In other words, if one of his neighbors has a much higher payoff than the others, then the chances are good that he will imitate that strategy.

In a mathematical kingdom governed by this update rule, we can study the evolution of cooperators versus defectors. Cooperators pay a cost c for each neighbor to receive a benefit b. Defectors, on the other hand, pay no cost and distribute no benefits. Over many rounds of evolution, we study the abundance of cooperators and defectors.

We varied the benefit-to-cost ratio, b/c, and found that for increasing ratios, cooperators became more and more abundant. There was a critical benefit-to-cost ratio at which cooperators were exactly as abundant as defectors. If the ratio was lower than this critical value, then defectors won. If it was higher, then the cooperators emerged victorious.

There was a simple link between cooperation and the structure of a network. Overall, the game was easier for cooperators if each individual had fewer neighbors. The average number of neighbors is called the degree of the graph, k. For example, a cycle has a degree two, because each individual has two neighbors. What took us aback was that the computer simulations seemed to suggest the following simple rule held sway over all kinds of networks: if the benefit-to-cost ratio is greater than the degree of the graph, then cooperators are more abundant than defectors. We were surprised and excited that such an elegant rule might hold.

We asked Hisashi if he could take this conjecture, which was based on numerical simulations, and generate a mathematical proof. It took him a long time by his standards. But in the end, he succeeded. He had a proof for this conjecture. Now we knew it was true. I was amazed and overjoyed. It was mind-blowing that such a simple rule (expressed as if $b/c > k$, then cooperators outnumber defectors) could be true and, just as remarkable, that it had never been observed before.

As I had mentioned at the start of this chapter, defectors always beat cooperators when they encounter each other in a well-mixed population. But on a graph, cooperation can thrive when cooperators huddle together to form clusters. From Hisashi's rule we can see that it is easier to form a cluster if each individual is only connected to a few others. The fewer the neighbors, k, means the smaller the benefit-to-cost ratio that is required for cooperation to thrive.

Once again, the update rule is very important because it specifies how players learn from each other. There are many different plausible update rules and it turns out that any given population structure can support evolution of cooperation for certain update rules but not for others. Cooperation can emerge if the update rule is extrovert and says:

Which of my friends is doing well? Is he a cooperator or a defector? If the former, then cooperate. However, cooperation cannot thrive if the update rule is introvert and says: I compare myself with a friend; if I am doing better I stay with my strategy; if my friend is doing better I adopt her strategy.

The reason for the difference resonates with the example I used at the start of the chapter. Let's use an outgoing update strategy. I want to learn from my friends who seem to be fashionable, whether in terms of what they wear or the music they like to listen to. I look at what they like, then buy the same clothes and download the same tracks. This leads to cooperation. Now let's use a myopic and self-centered update rule. What has made me successful so far? Choosing these particular clothes and songs. I decide to adopt this strategy, no matter what. An inevitable consequence of this update rule is that I erode cooperation in my network.

In broad terms, this work has raised the fascinating idea that certain structures of social network promote cooperative behavior better than others, particularly if the connectivity between players is low. One can easily see this in real life. With four of us working on games on graphs, our motives were strongly interlinked. If there had been forty, the overall effort would have been harder to manage and would have lacked spontaneity. One can see how businesses could use this kind of analysis to work out the ideal size of a dream team, building on earlier work, such as one study that revealed the ideal balance of rookies and veterans. Now we can design the optimum structure for cooperation in corporate institutions, one that promotes cooperation.

Examples of games on graphs abound in everyday life. Even if we know many people, we have only a few close friends. Those close friends are the ones we trust enough to engage with them in a cooperative dilemma (for example, to share a holiday villa, or to write a joint book). This network of close friends has the propensity to promote cooperation even in the absence of the conditional strategies of direct and indirect reciprocity. But when the effects of reciprocity are taken together with those of such structures linking a few close friends, there is a synergistic effect far and away beyond what smart players can achieve in a well-mixed population.

In the same way, the "if $b/c > k$ then cooperators outnumber defectors" rule shows that the fewer friends you have, the more strongly your fate is bound to theirs. There were three Musketeers, not thirty. There were the Magnificent Seven, rather than Seventy. I find it reassuring that we can connect the arcane world of mathematics with an issue that is as human and real as friendship. It is even more reassuring that math can say something precise about human cooperation that backs what we know from common experience. When the going gets tough, help is likeliest to come from those closest to you.

CHAPTER 13

Game, Set, and Match

It's not what you know but who you know.
—Proverb

Groucho Marx once sent the following telegram to the Friars Club of Beverly Hills: PLEASE ACCEPT MY RESIGNATION. I DON'T WANT TO BELONG TO ANY CLUB THAT WILL ACCEPT PEOPLE LIKE ME AS A MEMBER. In fact we all belong to clubs. Even Groucho. Society is a vast and sprawling multidimensional tapestry of clubs. They don't have to depend on formalities, such as having to wear the right tie. They can be allegiances. They can be friendships too. Or simply sets of people with aligned interests. Membership in the same organization becomes a good reason for friendship to bloom between two people and for a link to be forged between their respective social networks.

Perhaps the origins of that new link lie in the past, in the membership of a club. You may have come across that person before, when you studied at kindergarten, school, or university. Perhaps you both shared a formative experience. You may feel closer to people who, like you, celebrated a triumph in the same sports team or celebrated the same set of finals. You may have both survived a terrifying ordeal, such as serious illness, a car crash, or a bomb attack.

Or perhaps the link between you resides in the present. You support the same cricket team. You both like the mouth-burning qualities of vindaloo curries. Or you share the same environment, such as working at a particular kind of job. It could simply be that you are both very rich. As an impoverished Austrian comedian long ago once asked, Why is it that millionaires only invite other millionaires for dinner?

We find it easier to befriend someone who belongs to the same clique, whether he is an alumnus of the same college, supports the same baseball team, sends his kids to the same school, and so on. If you meet another person who belongs to several of the same sets (fan of the Boston Red Sox, follower of heavy metal music, and devotee of modern art installations), then the chances are that you already feel you know each other to some extent.

To find how much one has in common with another person is one of the first thrills of making a friend and of falling in love. And there are circumstances where, even when we don't have much in common with another person, we look up to her and will do anything to join her particular club of followers—change our hairstyle, follow another soccer club, or whatever to more easily fit in with her entourage. We all want to hang out with cool people, after all. We all hanker to be hip. We all want to be members of the smart set.

Because of our ever-changing tastes and allegiances, our networks of contacts are complex and in constant flux. A link in your social network can be sundered when you leave a club, whether a place of work, synagogue, or street. Or when you decide to support a different football team. Or when a friend loses all his money and has to sell up and move out to a cheaper neighborhood. Or when your partner meets someone she finds wittier or more attractive.

How does the membership of the sets that we belong to—and these myriad social networks—affect cooperation? Of course, it makes sense that if I meet you in several social sets, then we have more opportunities to interact. But to what extent do I start to cooperate more? Does it help if I join more sets, since that means I will spend less time in each one? Does it help if I join smaller sets, since their size should make it easier to meet other people?

As ever, I wanted to view this fundamental issue through the lens of evolutionary thinking in what became evolutionary set theory. The main question turned into, How can we understand evolution dynamics in populations that are structured around sets?

While the earlier work on graphs assumed a static population structure, evolutionary set theory had the power to capture the effects of fluidity and change as, for example, people flit between sets. This could offer powerful new insights.

A NEW DAWN

No man is an island . . .
—John Donne

The first answers to these questions were to come from Corina Tarnita, a virtuoso mathematician with a genuine interest in understanding the world in which we live. She defies the unfair stereotype that mathematicians are geeky and gauche. Equally, she confirms the stereotype that the very best in her field are young and enthusiastic. She started out as a pure mathematician, engaged in studying deep theories and finding new mathematical insights for their own sake, because they were aesthetically beautiful, rather than searching for any practical application. Her work on sets would bridge the Platonic world of mathematics and the very structure of human society.

Raised in the city of Craiova, in southwest Romania, Corina was tested and challenged from the very outset. Her mother, a professor of physics and materials science, would set her puzzle after puzzle from the age of three onward. "Everything was about maths," Corina recalls.

A prodigy, Corina participated in the National Mathematics Olympiad from the age of twelve. Most participants find the Olympiads daunting, but for Corina "they were fun." In three consecutive years she would come to win the national contest, where "mathletes" compete in a brutal process of elimination tests. By the age of eighteen this Olympic champion wrote her very first book on mathematics, to help prepare successive generations of bright-eyed students for the punishing intellectual rigor of the competition.

The same year, Corina was accepted by Harvard to study math, itself a huge achievement. The department there is one of the very best in the world, populated by many brilliant minds and stellar mathematicians. One of them would be a reader of Corina's undergraduate thesis, Shing-Tung Yau, who was brought up in poverty outside Hong Kong and rose to become a Fields medalist (the math equivalent of the Nobel Prize). He became the "emperor of math" in his homeland of China and inventor of mathematical structures known as Calabi-Yau manifolds that feature in string theory, the supposed "theory of everything."

One measure of the heady and challenging atmosphere at the Harvard math department is that it is governed by some surprising rules. Undergraduate students are always banished to do their graduate studies elsewhere. Likewise their own assistant professors cannot immediately be elevated to become full professors. They too must be exiled. The overall aim, always, is to keep fresh intellectual blood, and thus a steady stream of novel ideas, pumping through departmental veins.

When Corina's time as an undergraduate student came to an end, she became the exception to the rule. Corina had done her undergraduate thesis under the supervision of Joe Harris, a great algebraic geometer, who lobbied for her stay and invited her to join his team. She began to work in his field, which, as the name suggests, combines techniques of abstract algebra with the problems of geometry. To make progress, only pencil and paper and a trash can were required along with a very fine mind. Like many areas of mathematics, it took a huge effort simply to get acquainted with the field.

A strange twist of fate would send Corina my way. She had gone to the math library and started browsing the shelves, where she came across a copy of *Evolutionary Dynamics*. She started to leaf through my book and became intrigued by what she read. Corina had encountered evolutionary biology in Steve Pinker's lectures but was surprised to find that I had sketched out evolutionary ideas in a mathematical form. When we first met, I realized that she had a mind that was both mathematical and fascinated by biology. That is a rare combination.

She decided to step down from the top of the ivory tower of pure mathematics into the dark and tangled undergrowths of biology. To get started, I suggested she talk to others in my group to learn some of the techniques that we used. Like any tribe of science, we had our own particular customs, rituals, and techniques. For three hours, she interrogated Hisashi Ohtsuki. He has the knack of being able to express complex ideas with laserlike precision and was delighted to teach her. Even so, he had to ask her to come back the next day, when he gave Corina another long lecture. Hisashi wanted Corina to know everything that he knew. She also spent many hours with Tibor Antal, our Hungarian physics wizard.

I was amused by her relentless drive to absorb everything and anything of value before attempting her first problem. It struck me as a very feminine strategy. Male arrogance might demand to solve the problem first and ask questions later. The good news was that, after many intensive hours of discussions with members of my group, Corina decided that she had found her calling. She turned down a lucrative offer from a hedge fund in New York and told me she wanted to join my team. The die was cast.

She arrived at precisely the right time. She wanted to tackle a Big Problem and I had the perfect one for her. A few weeks before, I had been electrified by an idea that came in the form of an abstract image: some intersecting oval curves with dots scattered around in them. I realized that in this image lay an entire new way to think about population structure and evolution.

This was a new way to model populations that seemed entirely right for describing the fluid interactions created when one creature deals with many others. In fact, it looked perfect for the particular job of examining animal societies, such as the pecking order of apes and chimpanzees, as well as for human society, where all 7 billion people on the planet belong to one set or another. I decided to christen this new approach "evolutionary set theory."

But there was a problem. After a few days, I forgot the details of how, precisely, I planned to attack the problem. I had a vivid recall of the blaze of happiness that came from my discovery, but not the underlying reason for it. I nervously flicked through my notebook. On this particular subject, it was frustratingly blank. I was worried that I had lost this creative little spark (like so many others). I resolved not to think about anything else until I had recovered the idea. I sat down and, with some effort, managed to dredge up the details from deep in my memory. There it was, the outline of evolutionary set theory. The field grew out of a simple question: How do we study evolutionary dynamics if the members of a population belong to different sets?

We all belong to clubs, sets, and groups of one sort or another. In Britain, for example, journalists could work for the *Daily Telegraph,* the *Times,* or the *Independent.* But the reality is even more complex than this. You could write for the *Telegraph* and belong to a group that has written science books. Or to a set of people who live in the same street as you. Or to a set that likes raspberry jelly. Or one that hates thick-cut marmalade. Everyone belongs to more than one set, from the place where they work to where they work out, and this complicates how they will deal with each other and decisions they make about which group they belong to.

The structure of human society can be described by set memberships. You are more likely to meet and interact with people who belong to your sets. If you share several sets in common with somebody, then you are even more likely to interact with that person and you are even more likely to share mutual interests. That has to be fertile ground when it comes to understanding how people meet and why they cooperate.

How does thinking about set membership affect successive encounters between players in a game such as the Prisoner's Dilemma? Players interact with others who share the same sets. And, for the game to be realistic, these set memberships have to be fluid—they have to vary and change as much as they do in real life. So if you decided to back a new soccer team, you might find that your set of friends changed accordingly to include more supporters of that team. If you joined a new tennis club or changed jobs, you will start to interact with new people.

Evolutionary graph theory, which we encountered in the last chapter, deals with a snapshot of a network in a population. There we studied how a given, fixed population structure affects the outcome of evolution. The same goes for spatial games we encountered in chapter 3. In the new theory, the set memberships imply a network at any given moment in time. However, that network changes as people change their allegiances and move about between sets. Graphs are solid; in contrast, evolutionary set theory is fluid, just like human relationships, which, with all their foibles, are subject to the ebb and flow of many influences.

One of the most powerful is the aspiration to learn from successful people. We want to imitate what they do, wear their clothes, and join their clubs. It is not difficult to think of examples. If other researchers saw Corina's work and thought it was innovative and cool, then they might want to work on this topic, perhaps even join our team. The most important idea is that the concept of games on sets leads quite naturally to dynamical—changing—graphs. If we could capture this in the language of math, we could chart in detail how interactions wax and wane over time, as alliances are forged and sundered.

There was another advantage to looking at populations this way. While networks capture some aspects of our relationships, sets capture many more. This is easy to appreciate: When I am connected to two people who are on a given social network, I do not necessarily know if these two are also directly connected. But if I find myself in the same set with two other people, then I know—as they do—that we all belong to that set. Set membership makes common interests public, like the lists of sponsors of an art gallery or the hash tags used in Twitter.

Charting the evolution of the sets that we belong to held a deep appeal for a young mathematician like Corina, whose professional interests were divided up among subjects as diverse as algebraic geometry, number theory, theoretical computer science, cognitive psychology, and behavioral economics, and whose personal interests stretched from fast cars to music and good conversation. She thought that this problem would help her apply her mathematical skills to understanding the living and messy world. Hungry for a challenge, Corina prepared for her assault on this daunting problem.

Although used to pencil and paper, she decided to use mathematical software so she could work more quickly and share her ideas more easily. It took her only a few days to get acquainted with the necessary programming. Then came yet more discussions before she got down to business. Tibor, in particular, played a key role in helping Corina to prepare to launch her first foray. After more thought and more deliberation and more discussion, she declared that she thought the problem could be solved analytically. I was skeptical. But it struck me that inexperience has one great benefit. It makes us brave (or foolhardy, depending on the outcome).

After a few weeks there were no clear results. She encountered one difficulty after another. From time to time, Corina seemed to be making great progress but then, once she had climbed atop one peak, it turned out to be the shoulder of yet another, which was bigger still. In mathematics, that often spells the end of one particular intellectual line of ascent, a dead end and a return to base camp, cold, demoralized, and disappointed. Doubts set in. I began to worry.

At one point, the prospect of her ever making a breakthrough seemed to evaporate altogether. One of Corina's key supporters had been the great Hisashi. But he decided to leave our group so that he and his wife, Akiko, could return home to Japan. When Hisashi walked out of our team, it was a big loss for all of us. His power of analysis and his clarity of thought are peerless. He was our Sherpa Tenzing. We needed another guide in the ascent. When Corina temporarily became bogged down in the foothills of her enormous calculation, Tibor Antal took over from Hisashi to help her to struggle upward to the next summit. Together, they would work out the best route and figure out the mathematical equivalent of crampons, ropes, and ice axes to help ascend it.

Tibor had showed up in our group a few years earlier. He was a regular attendee of our seminars and always asked the best question of all those present. That was a good sign. When I first encountered him, Tibor worked at Boston University, which sprawled along Commonwealth Avenue on the other side of the Charles River. He was keen to join our group and I was only too happy to welcome him on board. He has a refreshing attitude too. He knows how to live. A lover of jazz and beer, he often drags us out to gigs and for a drink.

Tibor had a huge thirst for problems too and, even better, came from a different problem-solving tradition. He takes an approach that is typical of a physicist, not a mathematician. Physicists tackle unsolvable problems because they are content with approximate solutions. They do not let the quest for the best drive out the good. A brilliant exponent of this pragmatic approach was my mentor Bob May. But, although this strategy gets results, it can furrow the high brows of purists. Mathematicians yearn for exactness. Richard Taylor, the great Harvard number theorist, once asked me, "Do you have a mathematical proof for this statement or only a physicist's plausibility argument?" I loved the way Taylor formulated his question. I immediately replied, "Only a physicist's plausibility argument," to steer out of troubled waters.

The good news for Corina was that Tibor brought with him some useful mathematical climbing equipment, developed in earlier work on games in "phenotype space," in which players might behave differently toward those who look similar or dissimilar. But Corina still had to invent new techniques and concepts to overcome this problem. This is where Tibor's character mattered as much as his math. Throughout, he had repeated the same advice to Corina: Don't give up. Keep going! And so she did. Then, one day, the summit seemed to be in sight.

After four months of grueling ascent, with many setbacks, she conquered the problem of evolutionary sets. She had a formula. It was not the simple kind (like $E=mc^2$ or *If b/c > k, then cooperation thrives*) but a lumbering monster, a forest of symbols. She claimed that this leviathan, which filled several pages, represented the *exact* solution to the problem. It was a mathematician's formula, not a physicist's approximation. I wanted it to be true but I was skeptical.

I ran computer simulations to check her workings. Even though I had written an efficient program, I had to leave my computer to chug away overnight. The next morning, the results appeared. There was a perfect match between her formula and the data. I had never seen such an agreement in a calculation of a complex biological system. Corina's formula gave the exact answer. She had indeed conquered the problem. It was game, set, and match.

TRENDSETTING

From the lofty summit of this mathematical achievement, we had a commanding view of evolutionary dynamics. We could reveal the precise conditions for natural selection to favor cooperation over defection in populations that are structured around sets. One simple conclusion that we could draw from Corina's work was that the more sets there are, the better it is for cooperation. The reason is that when there are more sets, cooperators have more opportunities to escape. It is easier for them to find sets that are free of the troublesome defectors that could exploit them.

Another twist of the model that provides a powerful engine for the evolution of cooperation is that individuals might only begin to interact with each other if they have several sets in common. When I realize that the person who has joined my tennis club also studies theoretical biology, for example, then I will be more likely to begin collaborating with her. By the same token, it is not enough that we are both Democrats, or that we both shop at the same supermarket, or that we live in the same neighborhood. To have a reasonable chance of cooperating, we have to be Democrats who are neighbors and shop at the same supermarket. This "choosiness" of cooperators dramatically enhances their probability of success. As a result of this feature, sets are the structures with the greatest potential to promote the evolution of cooperation.

Corina's equation makes a fascinating prediction. It suggests that there is an optimum level of mobility (meaning the rate at which people move between different sets and explore new sets). If the mobility is too low, then the population is too static and cooperators can be exploited by defectors. But the converse is bad for cooperation too. If the mobility is too high, then any "fellowship of cooperators" that fosters mutual help does not persist for very long. The most fertile ground for cooperation comes between these two extremes.

With intermediate mobility, cooperators have a chance to hang around for long enough to benefit each other, but they can also escape from defection by colonizing new sets. The process is guided by natural selection: if a few cooperators find a new set without any defectors then they do well and attract more cooperators. Only after some time one of them might switch to defection and, by doing this, destroy the happy situation that had once thrived there. Then the resident cooperators have to find a new set. Because it is harder for sets with defectors to attract new members, over time they dwindle and, eventually, become empty.

COOPERATIVE DILEMMAS

In this and earlier chapters we have encountered various ways in which population structure can promote the evolution of cooperation. We saw how the dark forces of defection can be defied in spatial games, in games on graphs, games on sets, and also when there is competition not only between individuals but also between groups of individuals (so-called multilevel selection). Is there a deeper concept that underpins all these apparently different approaches to cooperation? Is there a simple rule that might hold sway for all of those cases? Surprisingly, it turns out that there is.

In order to understand this simple rule, let us step back and once again consider a basic game, one between two people. Each person can choose to behave in one of two ways. As ever, we will divide them into being either cooperators or defectors with the usual payoffs: R = reward for mutual cooperation; P = punishment for mutual defection; S = sucker's payoff; and T = temptation to defect. And, once again, we will arrange these payoffs so that T is greater than R which is greater than P which is greater than S, so that we end up with the Prisoner's Dilemma. As I mentioned in the opening chapter on the Dilemma, this ranking of payoffs ensures that we have the toughest form of the cooperative dilemma.

Overall, a cooperative dilemma is defined by R being greater than P—in other words mutual cooperation is greater than mutual defection—and at least one of the following temptations to defect: T is greater than R; P is greater than S; or T is greater than S. When T is greater than R, it means that if the other person cooperates it is better for me to defect; P greater than S means that if the other person defects, it is better for me to defect too; and T greater than S means that in a pair formed of one cooperator and one defector, it is better for me to be the defector. If none of the three incentives to defect are present, then the game is not a cooperative dilemma. In this case "cooperation" is just the best thing to do. It is a no-brainer.

In his explorations of this world of cooperative dilemmas, Tibor Antal had come up with a beautiful result. Let us consider a well-mixed population, where every player is equally likely to interact with every other. Individuals play the game, accumulate payoff, and tend to imitate the strategy of other successful players. Hence there is natural selection between the two strategies that is proportional to payoff. But let us also introduce mutation between the two strategies, which means people can sometimes switch from cooperation to defection at random. Tibor showed that cooperators are on average more abundant than defectors if $R + S$ is greater than $T + P$.

What does this condition tell us? $R + S$ is simply the average payoff that a cooperator receives if he is equally likely to encounter cooperators or defectors. Similarly, $T + P$ is the average payoff that a defector receives if he is equally likely to encounter cooperators or defectors. (In both cases we have omitted the factor ½ because it cancels out.) The condition "$R + S$ is greater than $T + P$" means that the average payoff of a cooperator is greater than that of a defector. This inequality could never hold true for the Prisoner's Dilemma. In a well-mixed population, if all players are in the grip of this Dilemma, then cooperators are always worse off than defectors. But for other cooperative dilemmas, the condition might hold. Then it might pay to cooperate, even in a well-mixed population.

GEOMETRY OF COOPERATION

Tibor Antal's elegant result holds true for well-mixed populations where any two individuals have an equally likely opportunity to interact. But can we find a similar result that would hold true for structured populations? Keep in mind that structured populations come in an infinitude of shapes and flavors, while the well-mixed population is only one example, and a very particular one at that. To have a general statement that reigns over all structured populations would be amazing.

Yet, over the years, I have collected results that provide a tantalizing hint that this remarkable feat should be possible. For many different models I found that the question of whether natural selection favors cooperators over defectors can be answered by a simple variant of Tibor's formula. The variant is simple, because it is down to the addition of one single parameter called the structure coefficient, which I have denoted "*sigma.*"

This coefficient specifies the relative rate at which like-minded players meet: in other words, the relative rate at which cooperators team up with other cooperators, and defectors gang up with other defectors. The average payoff of a cooperator becomes *sigma* times *R* plus *S*. In a similar way, the average payoff of a defector becomes *T* plus *sigma* times *P*. As before if the average payoff of cooperators is greater than that of defectors, then cooperators tend to be more abundant than defectors. Therefore, whether cooperators are victorious over defectors depends not only on the payoff values (*R, S, T, P*) but also on the value of *sigma*. If *sigma* is greater than one, then cooperators are even able to win in a Prisoner's Dilemma.

I found that every model that we had studied—however sophisticated—could always be reduced to such a simple linear inequality in the case of weak selection. That meant that some essence of each population structure, no matter how complex, could be captured in the value of its *sigma* parameter. But, of course, calculating this structure coefficient, *sigma,* for any given model is where the real problem lies. It turns out that when Corina "solved" how to play games on sets, she had in effect found a way to calculate *sigma* for sets.

The way this parameter works is easy to understand. If *sigma* is greater than one, then like associates with like: what we call positive assortment or clustering. If *sigma* is less than one, then the opposite strategies interact more frequently. Hence there is negative assortment. For the well-mixed population *sigma* is equal to one. So if we want cooperators to thrive in the Prisoner's Dilemma we need positive assortment and thus need *sigma* to exceed one. Corina calls this effect "divine Tit for Tat"—if you are cooperator, you find yourself surrounded with cooperators and vice versa. Or to put it another way, what you reap is what you sow.

Over the years I had been hoarding such *sigma* values, like a naturalist acquiring beetles, from big to small, from brown to iridescent. But when I discussed my diverse and magnificent collection with Corina, she began to wonder whether she could actually provide a mathematical proof that every population structure must always lead to a simple mathematical expression with a single structure coefficient. After some time, this is precisely what she achieved. Her proof was breathtaking. Even for a seasoned mathlete, overcoming "every population structure" was like conquering every peak in a vast and rugged mountain range.

By thinking about something as intimate as social relationships, we have come to a cosmic conclusion. In evolving populations, cooperation withers, shrivels, and dies when the structure coefficient, *sigma,* is less than one. Equally, when *sigma* is greater than one, cooperation takes root and flourishes. Corina's theorem will hold for any evolutionary process on Earth, in this galaxy, as well as all the others, from those nearby to agglomerations of ancient stars that lurk in the faintest, farthest reaches. It applies to any and every game in the cosmos.

CHAPTER 14

Crescendo of Cooperation

Die liebe Erde allüberall
Blüht auf im Lenz und grünt aufs neu!
Allüberall und ewig blauen licht die Fernen,
Ewig . . . ewig . . .
—Mahler, *Das Lied von der Erde*

"Imagine a work so large that it mirrors the entire world." With this evocative sentence, Gustav Mahler summed up his ambition to create a new kind of music. The Austrian composer wanted to write symphonic works of such remarkable scope and magnitude that they could summon the fundamental forces that created the cosmos from the void. True to his word, Mahler's compositions are epic and heartstopping. They are about life, death, love, and redemption. They are universal statements about the human condition, from its highest, brightest glory to its lowest, blackest folly.

Mahler took his own struggles with the vicissitudes of life and channeled these visceral experiences, along with his hopes and fears, into monumental orchestral works. He used his music to wrestle with the primeval moment of creation, even with the fundamental forces of evolution itself. He hoped that in some way he himself would become an instrument that could be played by the whole universe. I find his music, its huge breadth and its all-encompassing ambition, a tremendous and enduring inspiration.

Of all his works, one of the most striking is Symphony no. 8, which was dedicated to his wife, Alma. A work of reconciliation, the symphony celebrates the redemptive power of love. The première took place in Munich on September 12, 1910, and featured a chorus of about 850, with an orchestra of 171. In recognition of its epic scale, Mahler's agent nicknamed the work *Symphony of a Thousand.* Even today, the logistical demands of this work put it beyond the regular run of concert life and make any airing of it a major event, a striking tribute to creativity and cooperation as a battalion of musicians explore its complexity, its overwhelming intensity, and its sheer expressive joy. The opening theme of this grandest of all symphonies was articulated to fit the words *Veni, creator spiritus*—"Come, Creator Spirit." The second part is the apotheosis of Goethe's Faust.

Over the past decades I have voyaged through distant and diverse areas of science to seek out what I believe is the most creative force of biology, the one we know as cooperation. It is manifested at every level of human society, from an orderly queue of strangers at a railway station to the organization of a rock concert at a Super Bowl. The degree to which we cooperate sets us apart from the rest of creation. This is the fundamental reason humans have managed to eke out a living in almost every ecosystem on Earth and indeed have started to venture well beyond Earth. But, of course, this raises all kinds of questions, which I have examined in earlier chapters—not least, the one that troubled Darwin himself: In the ceaseless competition for food, territory, and mates, why would one individual go out of its way to help another?

Cooperating with many dazzling people over two decades, I have studied various ways in which evolution leads to cooperation in our highly competitive world. The basic issue that we have explored can be couched in terms of cost and benefit. A cooperator pays a cost for another individual to receive a benefit. If the cost is larger than the benefit, then cooperation is not productive and the game is not a cooperative dilemma. In this case, two cooperators would be worse off than two defectors. But if the benefit is larger than the cost, then we end up with a familiar game, the Prisoner's Dilemma.

Here is the problem that is central to the Prisoner's Dilemma. In the simplest version of the Dilemma, without making any additional assumptions, natural selection favors defectors. As mentioned before, cooperators always have a lower fitness than defectors in a well-mixed population. As a consequence, as that population evolves, natural selection slowly increases the abundance of defectors until every last one of the cooperators has been exterminated. This is the "wrong" outcome, because a population of cooperators has a higher productivity (higher average fitness) than a population of defectors. Hence in this particular case natural selection does not achieve the highest fitness but actually destroys what would be best for the entire population. To favor cooperation, natural selection needs help. It needs mechanisms for the evolution of cooperation.

At present, we know five mechanisms for the evolution of cooperation. I have studied how these mechanisms work by blending game theory with evolution, assuming that the payoff from a game affects reproductive success. This means that, as the players mutate and evolve, natural selection smiles on players that have a high payoff. They reproduce relatively more in the struggle for existence, while fellow players that are unsuccessful dwindle and then die off.

I do not restrict the use of the term "natural selection" to genes alone. Depending on whether we talk about cells, animals, or people, reproduction can be genetic or cultural. In the former case, successful individuals leave more offspring and pass more genes on to future generations. In the latter, successful ideas, fashions, and strategies spread by imitation and learning: a fad is born. As one example, the concept of Darwinian evolution itself does not spread genetically but culturally, leaping from the mind of one biologist to infect another.

My work shows how cooperation arises out of competition, even though the two are locked together in ceaseless conflict. The collective effort of society depends in part on suppressing the ability of the individual to mutiny and defect. The same goes for rebellious cells, chromosomes, and genes. Like day and night, or good and bad, cooperation and competition are forever entwined in a tight embrace.

MECHANICS OF COOPERATION

To reap the rewards of cooperation there has to be at least one mechanism at work to counter the relentless and depressing tendency of natural selection to grind down the average fitness of a population in the Prisoner's Dilemma. In the opening chapters, I described five such mechanisms and how they can make us work together:

1. *Repetition* (direct reciprocity), "I'll scratch your back and you scratch mine." This accounts for the success of Tit for Tat–like strategies, whether those of the little fish that offer cleaning services on coral reefs, generous vampire bats that share blood meals, or the military units that policed the unofficial truces that emerged on the Western Front in the First World War, where an accidental infraction was met with a revenge raid or barrage. As the great eighteenth-century Scottish philosopher David Hume wrote in *A Treatise of Human Nature* in 1740: "I learn to do service to another, without bearing him any real kindness: because I foresee, that he will return my service, in expectation of another of the same kind."

   We also saw how, when there are mistakes caused by trembling hands and fuzzy brains, it is better to depend on strategies such as Generous Tit for Tat or Win Stay, Lose Shift. Relative to the former, the latter is the even simpler idea of repeating your previous move whenever you are doing well, and changing when the going gets tough. I described how, overall, direct reciprocity can lead to the evolution of cooperation only if the probability of another encounter between the same two individuals exceeds the cost-to-benefit ratio of the altruistic act; that's our first simple rule.

2. *Reputation* (indirect reciprocity). This mechanism of cooperation thrives when there are repeated encounters within a group of players. Now my behavior toward you also depends on what you have done to others. To get any return from being nice to someone, we must place our faith in forthcoming encounters: "Give and it shall be given unto you," as Luke put it in the Bible. One can sum up this mechanism with the phrase "I scratch your back and someone will scratch mine."

In human society, indirect reciprocity relies to a great extent on communication. In chapter 2, I explained how language might be needed to learn from the experiences of others and thereby establish the reputation of people, as well as pass it on again. We found that indirect reciprocity can only promote cooperation if the probability of knowing someone's reputation exceeds the cost-to-benefit ratio of the altruistic act; that's our second rule.

3. *Spatial selection.* This process occurs on the chessboard of life or in the spider's webs of social networks or the myriad sets that we all belong to. At the heart of any evolutionary process is a population of reproducing individuals, and the work of many scholars over the years has shown how the structure of that population can affect evolution. Whether we are talking about spatial structures or social networks or tags, all we mean by this is that some individuals interact with each other more often than others. Now cooperators can prevail by forming networks and clusters in which they help each other. Just as a gravitational lens bends the light of a galaxy, so the structure of a population bends the trajectory of evolution. A surprisingly simple rule determines whether cooperation can bud and flower on graphs. The benefit-to-cost ratio must exceed the average number of neighbors per individual. That's our third rule.

4. *Multilevel selection.* This mechanism recognizes how, in some circumstances, selection acts not only on individuals but also on groups. A group of cooperators might be more successful than a group of defectors. No one sums up this mechanism of cooperation better than Darwin: "There can be no doubt that a tribe including many members who . . . were always ready to give aid to each other and to sacrifice themselves for the common good, would be victorious over other tribes; and this would be natural selection." Multilevel (group) selection allows the evolution of cooperation, provided that one thing holds good: the ratio of the benefits to cost is greater than one plus the ratio of group size to number of groups. Thus this cooperative mechanism works well if there are many small groups and not so well if there are a few large groups—our fourth rule.

5. *Kin selection.* Here the bonds of family and of common ancestry are

decisive. I recognize my kin and I behave accordingly, so that I cooperate with close kin and I defect with strangers. Another common way to express this drive to look after one's own, genetically speaking, is to say that blood is thicker than water. This is summarized by Hamilton's rule, which states that the coefficient of relatedness must exceed the cost-to-benefit ratio of the altruistic act; this is our fifth rule. Although I have discussed the problems of this approach in chapter 5, I still believe that kin selection is a valid mechanism if properly formulated.

There we have it. Using these five mechanisms of cooperation, natural selection has ensured that we are able to get more from social living than from the pursuit of a solitary, selfish life. Thanks to these mechanisms, the essentially competitive drive of evolution can, in many circumstances, give rise to cooperation. Because our instincts have been shaped in this way over the generations, it is no surprise that one corollary of this is that universal behaviors—such as love, friendship, jealousy, and team spirit—are seen across all human societies.

If cooperation thrives due to the mechanism of multilevel selection, for example, then although there might always be an incentive to defect, those groups who have a higher percentage of people willing to sacrifice themselves for the greater good can do better. A nation, cult, or a religion can be seen as a group that is bound by the way that an individual makes sacrifices to help his brethren.

In direct or indirect reciprocity, we can glimpse the traditional idea that "one good turn deserves another." What I find amazing is that these calculations show that despite nature's competitive setting—based on natural selection—the winning strategies of direct and indirect reciprocity must have the following "charitable" attributes: be hopeful, generous, and forgiving. Hopeful here means that if I meet a newcomer then I hope that I can establish the basis of cooperation with him by making an effort to cooperate. Forgiving means that if someone defects, then I will work hard to reestablish a relationship based on cooperation. Generous means that in most of my interactions with other people, I do not adopt a myopic perspective. I do not moan about who is doing better than me and who is getting the bigger share of the pie. Instead, I am content with equal or even slightly smaller shares but enjoy many productive and helpful interactions overall; now many more pies get shared.

In this way, my work on cooperation highlights which kinds of behaviors are important for human evolution and success in daily life. We have five mechanisms that can work separately and together to help everyone to get along. What is remarkable is that from an analytical, quantitative, and mathematical basis I can come up with ideas that should seem as familiar to secular ethicists as they are to followers of religions.

Diverse faiths are united by the reciprocity of the Golden Rule, as we saw in chapter 2. Evolution, which at first glance seems to present problems for faith, actually hones selfless, altruistic, and perhaps even saintly behavior. The teachings of the great world religions have much in common in that they provide ancient recipes for how to lead a fulfilled life. For millennia they have analyzed the human condition to ameliorate suffering and sadness. They have come to the conclusion that love, hope, and forgiveness are essential components of what is needed to solve the biggest problems. They call for unselfish action. Jesus says if you give, then your left hand should not know what your right hand is doing. Krishna says to the prince Arjuna in the Bhagavad Gita: You have to see yourself in every creature. You have to experience the sufferings of others as your own. For those who follow a faith, the solution comes when the drive to be selfish is overwhelmed by love. In the language we have encountered in this book, the teachings of world religions can be seen as recipes for cooperation. Now, for the first time, aspects of these powerful ideas have been quantified in experiments, captured in equations, and enshrined in science.

THE NEXT STEP FOR MANKIND

Mahler's Third Symphony chimes with my quest to understand the ultimate manifestation of cooperation, the 4-billion-year story of life on Earth. Written between 1893 and 1896, this is the composer's longest piece—a performance lasts almost two hours. The symphony is a pantheistic vision of the universe, a gigantic musical poem, a hymn to the natural world in the form of a step-by-step ascent of the great Ladder of Creation.

My love of this symphony dates back to my early years in Oxford, at the start of the 1990s. One day my new student, Sebastian Bonhoeffer, asked me if I would like to go to a concert in the Sheldonian Theatre, a seventeenth-century building by Christopher Wren that is one of the architectural jewels of Oxford. Sebastian, himself an excellent musician (and now a biology professor at the ETH in Zürich), took me to hear a performance of Mahler's Third, in which he played the lead cello. I came and I listened. It was my first encounter with the great composer. On the uncomfortable wooden seats of the Sheldonian, my whole life unfolded. I never felt the same way about music again.

Mahler begins with a slow, primeval opening that evokes inert matter—rocks and inanimate nature—and gradually accelerates so that it becomes rousing and pounding. Life then marches in. The symphony passes onward and upward through more elaborate stages of evolution—flowers, animals, and mankind—before reaching divine love, which Mahler imagined as a supremely transcendental force.

From the movement "What the Rocks Tell Me" to "What Love Tells Me," from the first awakening of antediluvian life to the final exultant moments of this huge work, Mahler hoped that "nature in its totality may ring and resound." He achieved his aim. On its first complete performance on Monday, June 9, 1902, in Krefeld, the work was received with thundering acclaim, marking a rare moment in Mahler's life.

One of his letters describes how in the symphony "Nature herself acquires a voice and tells secrets so profound that they are perhaps glimpsed only in dreams!" In my own way, I would like to think I have helped to give nature her voice too. I find it more nuanced and more subtle than the one revealed by reflecting on competition alone. I have argued that evolution "needs" cooperation if she is to construct new levels of organization, driving genes to collaborate in chromosomes, chromosomes to collaborate in genomes, genomes to collaborate in cells, cells to collaborate in more complex cells, complex cells to collaborate in bodies, and bodies to collaborate in societies.

After this grand tour of the mechanisms of cooperation, I have been struck—perhaps awestruck—by the extent to which humans cooperate. We use each and every mechanism that I have outlined, and to a remarkable degree. Sure, elements of every mechanism of cooperation can be glimpsed in other animal societies: multilevel selection among the ants, reciprocity among fish, spatial selection in colonies of bacteria, and so on. But no animal species can draw on the mechanisms to the same extent as seen in human society. Even our closest relatives, the apes, lack full-blown language and thus lack the full potential of indirect reciprocity.

While bacterial films may divide up labor across a single community, the complexity of this microbial cooperation pales in significance when compared to the complexity of human society that thrives in a modern city. While ants have a handful of castes, our society specializes to an extreme degree: from policemen and CEOs to soldiers to butchers, bakers, and candlestick makers. As Adam Smith recognized, using the example of a pin maker, the degree to which labor is divided up among the members of our society is extraordinary. What we learned in chapter 11 was that by rewarding successful cooperation, rather than by punishing defection, we can propel this cooperative effort to being truly creative and innovative. Working together this way, we can achieve things as a society that no individual ever could.

What makes us truly special is that cooperation through indirect reciprocity propelled the emergence of human language and brought about a new mode of evolution. We are now subject to an evolutionary dynamic that can detach itself to some degree from its genetic basis, from chemistry, genes, and DNA. This is cultural evolution, which involves learning, and explains why we are so devastatingly successful. As a result, the way the human brain evolves is utterly different from the evolution of any other biological structure that has ever existed. The architecture of the brain changes every time we talk to another person. We are able, in turn, to impose structural changes on the way the listener's brain is wired. The next time you listen to another person, remember that you have permanently changed the wiring of your brain and will do this every time you memorize a moment, no matter how fleeting.

Although this special ability makes human society breathtakingly cooperative, it can hardly be claimed that we live in a cooperative utopia. The last century saw hundreds of millions of deaths in civil conflict, world wars, and genocides. That effort to wage a war can be seen as a perverted form of cooperation. The people of one side join forces against the other in an organized effort that all too often leads to destruction. There is always antagonism and rivalry. There is always the danger that a new war will rage. Where there is cooperation, there is also the danger of exploitation. Defectors loom in the dark. Ready to strike. Waiting for the right opportunity to pounce and to take advantage. As shown since the very first simulations that I conducted decades ago with Karl, there are always oscillations. Cooperation comes and goes, waxes and wanes. It has to be reborn in endless cycles.

Today, mankind is teetering on the brink of several possible catastrophes of its own making. The danger of nuclear conflagration has not gone away but become so familiar and unfashionable that, relative to the size of this threat, it is hardly remarked upon. The cold war stockpiles of vast numbers of warheads persist to a significant extent. We still face a doomsday scenario where nuclear war—one started deliberately or accidentally—could throw so much dust, debris, and smoke into the air that it would block life-giving sunlight from our atmosphere, causing a "nuclear winter." In June 2005, Senator Richard Lugar, then the Republican chair of the Senate Foreign Relations Committee, asked about the prospect of a nuclear attack within the next decade. The seventy-six nuclear security experts he polled came up with an average probability of 29 percent. Four respondents estimated the risk at 100 percent, while only one estimated it at zero. As nuclear weapons continue to proliferate and as terrorism becomes increasingly organized, this danger has grown.

The recent slump in the global economy has given us a glimpse of what might happen if there was a collapse. Existing economic policies are based on the theory that the world is made up of a patchwork of simple, largely separate markets. Yet money can now flow more easily from country to country. This has stimulated global trade and prosperity, but it also means that an upset in one place can have major and unpredictable consequences elsewhere. The rapid increase in cross-border investments in recent decades is what allowed a local shock—the collapse in inflated U.S. real estate values—to propagate globally in 2008, especially through highly indebted investment companies that can respond to a loss of money in one place by withdrawing credit anywhere. In the event of financial Armageddon, world economies would slide into a punishing depression, perhaps worse than the Great Depression of the 1930s, when millions starved.

We are also staring into the abyss of environmental catastrophe, the ultimate Tragedy of the Commons. The Earth has a fever. And that fever is rising. The signs of climate crisis are now clear, from the drastic disappearance of the North Polar ice during the summer to faster-melting glaciers and the inundation of low-lying Pacific islands. There's a host of associated problems, from the water shortages facing cities in North and South America, Asia, and Australia to the spread of disease to problems with global food supply to an acceleration of the pace of extinction, as the web of life on which we depend is being torn, frayed, and shredded.

In his speech to accept the Nobel Peace Prize, Al Gore cited an African proverb that says, "If you want to go quickly, go alone. If you want to go far, go together." He gave a clarion call for cooperation. "We must abandon the conceit that individual, isolated, private actions are the answer. They can and do help. But they will not take us far enough without collective action." To put it another way, we are all prisoners of the same dilemma that ensnared myself and Karl years ago. And to solve apparently intractable issues such as climate change will take more than technology alone.

The danger is very real. I believe that intelligent life is fragile. I think that life has evolved in the universe often and has done so for the 13.7 billion years that our cosmos has been in existence. But, as far as we can see, we are alone. Intelligent life does not seem to stay around for long. This should give us pause for thought. Now, more than ever, we need to cooperate, and on a global scale. Although we are teetering on the brink of disaster, we are also on the brink of advancing to the next level of cooperation. We need a climate program beyond even the vast scale of the Manhattan Project, in which we mobilize our civilization with the kind of resource and resolve that have previously been seen only when nations mobilized for a world war.

In previous chapters, I described how new opportunities to cooperate are able to drive creativity. When competing units on one level of organization begin to cooperate, new and creative levels of organization evolved, first between molecules, then simple cells, complex cells, multicellular creatures such as people, and, finally, between societies. I believe that climate change will force us to enter a new chapter of cooperation.

At this very moment, people worldwide are linked together to an extent that is incredible by the standards of what came before. There are endless connections being forged and changed and sundered by the technology of modern communications, whether phones, mobiles, iPhones, BlackBerries, Androids, computers, or the web. At the same time, we now understand what it takes to work together better than ever. I hope that we can harness the new understanding of cooperation to rise to the challenges of our crowded, nuclear, feverish world.

THE ETERNAL SYMPHONY

Music is the pleasure the human mind experiences from counting without being aware that it is counting.

—Gottfried Leibniz

Though the ambition of the magnificent Third Symphony most closely matches the agenda of my research, the Mahler composition that I adore the most of all is *Das Lied von der Erde* (*Song of the Earth*). The inspiration for this work for voice and orchestra—a "song-symphony"—came from *The Chinese Flute,* a translation of ancient Chinese poetry by the German poet Hans Bethge. On reading Bethge's version, Mahler was moved by its vision of beauty, transience, and death. After completing the great orchestral work that this poetry inspired, he wrote that "it is probably the most personal composition I have created thus far." This was clear to his friend, conductor and composer Bruno Walter, when he first read the score.

When Walter saw how much of himself Mahler had poured into this transcendent work, he broke down and wept. The piece is suffused with a sense of mortality. This is hardly surprising. The summer before, in 1907, Mahler had been pushed, partly by anti-Semitism, to resign his post as director of the Vienna Court Opera, his eldest daughter Maria had died, and he himself was diagnosed with a serious heart defect.

In the second movement, Mahler refers to the death of his daughter ("a little light has gone out"), a shattering loss that he never came to terms with. The music tries to come to a resolution. It gets close but it can never quite make it. Like the Prisoner's Dilemma, it cannot ever be fully resolved. But, at the end of the last movement, Mahler finally becomes reconciled with his own death: "I will no longer travel into the distance. My heart is quiet and it is waiting for its hour." Reflecting on his own fate, he lets go.

Mahler was anxious about calling this work a symphony, fearing that like Beethoven and Bruckner before him, the ninth would be his last. Despite his unease, he dared to follow *Das Lied von der Erde* with a work entitled Symphony no. 9. Mahler had joked that he had cheated death as the new symphony was really his tenth. It turned out that his superstition would come true—the Ninth Symphony did prove to be his last completed work. A few months after Mahler's death in 1911, Bruno Walter led the first performance of *Das Lied von der Erde* in Munich.

A remarkable and famous reprise would be staged decades later, in 1952, when Walter returned to the piece with the Vienna Philharmonic Orchestra and the British singer Kathleen Ferrier, who had by then been diagnosed with breast cancer. The first time Ferrier performed the work she was overwhelmed by pain and raw emotion. She was unable to sing the final words of the final movement, "The Farewell." The music seems to evaporate into the ether, marking Mahler's final acceptance of death. The orchestra was touched and with Ferrier gave the performance of a lifetime. She lost her battle with cancer seventeen months later, at the age of forty-one.

All this may sound somber. But in the darkness of the symphony a chink of brilliant optimism can be glimpsed, along with a sense of surprise, which Mahler signals with a final change into the key of C major. At the moment that Mahler is reconciled with his own mortality, he understands how extinction will be followed by a new spring. This carries a deep resonance for me and my work.

Although *Das Lied von der Erde* underlines the fact that there's no escape for the individual from the blackness of death, life itself is reborn endlessly. As the music dwindles and dies away into nothingness, the teeming beauty of the world lives on. At the end, the earth is renewed in spring, everywhere and forever shining blue and bright. This final movement affects me deeply. Our world of change reflects an unchanging underlying reality that can be captured with mathematics. And just as the beauty of the world lives on beyond those final haunting moments of *Das Lied von der Erde,* so the laws of nature live on too.

The story of humanity is one that rests on the never-ending creative tension between the dark pursuit of selfish short-term interests and the shining example of striving toward collective long-term goals. I believe we now understand how defection in the Prisoner's Dilemma can be trumped by cooperation. And, just as Mahler ends on an upbeat note, so I believe the emphasis on cooperation puts a more optimistic sheen on life than the traditional take on Darwin, which condemns all life to a protracted and bloody struggle for survival and reproduction. Mutation and natural selection are not enough in themselves to understand life. You need cooperation too. Cooperation was the principle architect of 4 billion years of evolution. Cooperation built the first bacterial cells, then higher cells, then complex multicellular life and insect superorganisms. Finally cooperation constructed humanity.

I propose that "natural cooperation" be included as a fundamental principle to bolster those laid down by Darwin. Cooperation can draw living matter upward to higher levels of organization. It generates the possibility for greater diversity by new specializations, new niches, and new divisions of labor. Cooperation makes evolution constructive and open-ended.

Today we face a stark a choice: we can either move up to the next stage of evolutionary complexity, or we can go into decline, even become extinct. Though global problems loom large, we could be on the verge of the next transition in social organization, one of equal significance to the emergence of the first cell, of the first complex cell, or indeed of the very first multicellular creature. We have the understanding and, thanks to the remarkable extent to which our society is interconnected, we can build on it.

We are SuperCooperators. We are the only species on Earth that is able to draw on the support of all five mechanisms of cooperation, and we already do this to a remarkable extent. But we now have to do even better. We now have to strive to achieve the full potential of these mechanisms if we are to rise to the serious challenges that lie ahead.

Direct and indirect reciprocity will always play a critical role in routine dealings with others: the more help we give, the more we will receive. This aspect of human cooperation is as important as it is ancient. We are the only species that utilizes full-blown indirect reciprocity, because of our powerful and flexible language. When evaluating the reputation of another individual, animals must rely on direct observation. But, using language, humans can learn from the experiences of others. We can put a name to the face of someone and use it to create a reputation.

Reputation is a potent force that can be harnessed to avoid the Tragedy of the Commons. Success depends on freedom of information without censorship and spin. We need detailed information on the degree to which people, companies, or countries squander precious resources. We need to know the true environmental cost of everyday items, from a boiler to a car, so that we can build it into the price tag. We need to know the actual risks of climate change, without deviation, embellishment, or exaggeration.

Over the years we have spun ever expanding webs of indirect reciprocity from villages, to cities, states, and across the entire globe. Now, because of the high connectivity of our global networks, the reputation that is associated with a name can move around the planet in a matter of minutes. If someone has a great idea in Asia, a fellow scientist in the United States can learn about it instantly. If a thought-provoking blog is posted online, it can be disseminated, translated, and discussed worldwide that day. If a catchy new song is available for download, it can be played as easily in the main street of a little town as the back street of a vast urban ghetto.

Many of my own collaborations are with people who live on other continents. But because of email, Skype, phone, and so on, it is as if they are sitting in the room next door. My room in the woods of New England is as close to Cambridge as to Roger's home in London or to Hisashi's office in Tokyo. In this way, productive ideas and innovations can spread far and wide. Today, there are innumerable ways by which cooperation can flourish.

But with new opportunities come new dangers. All my work on the evolution of cooperation hints at one inevitable feature: there is no such thing as utopia and the degree of cooperation in a society will fall as inevitably as it will rise again. With globalization, the planet's resources are becoming exhausted. With globalization, the never-ending competitive quest to achieve economic growth is unsustainable. With globalization comes uniformity too, which makes us more vulnerable to jolts. As has been seen in the financial world, there is no longer safety in investing in diversified American, European, and Asian stocks. They are all interconnected, and when a financial collapse hits one market, they can all plunge into a nosedive. For the same reason, we are more vulnerable to pandemics: thanks to international air travel, a disease can quickly become established and spread around the world.

We cannot expect cooperation to endure forever. But we can hope to prevent a drastic fall, or at least ensure that cooperation is more likely to prevail over longer periods of time and only suffer the occasional breakdown. We can work to quickly reestablish cooperation after each collapse.

We need to place more faith in citizens than leaders. Cooperation has to come from the bottom up and not be imposed from the top down. That is why, for example, democracy is a cornerstone concept, since this is a form of cooperation that grows from the roots. We need to do even more to create an environment where cooperation can flourish, if we are to reap its creative benefits.

Another lesson of my analysis of the mechanisms of cooperation over the years is that we have to learn not to be too inward looking, petty minded, and competitive. When it comes to the structure of society, for example, we have to step out of the narrow confinement of looking after our relatives, or our own kind. Kin selection (even if properly formulated) is only a small component of human cooperation. Nepotism is counterproductive when it comes to cultivating cooperation across wider swaths of society.

We have to look beyond the narrow idea that punishment and threat can enforce cooperation. In my opinion, creative cooperation can only come from helpful interactions such as participation, friendship, and reward.

We also discovered how we need to be more open to absorbing the lessons that lie behind the success of other people, rather than focus on our own immediate goals. By adopting the former extrovert strategy, rather than the introvert latter, we can ensure that best practice will become established.

And, of course, we need to remember the legacy of Hardin. We need to find new ways of thinking—a fundamental extension of morality—if we are to live within our terrestrial means. And that brings me to a point that I have already made, but one worth repeating. Down the years my research has told me that if we are to solve the Prisoner's Dilemma, we need to be generous, hopeful, and forgiving. Perhaps for the first time, the conclusions of science and mathematics can be seen to intersect with the teachings of world religions.

Humans are SuperCooperators. We are able to draw on all five mechanisms of cooperation. In particular, we are the only species that can summon the full power of indirect reciprocity, thanks to our rich and flexible language. We have names and with them come reputations that can be used to help us all to work more closely together. We can also design our surroundings—from architecture to laws to the internet and much more besides—to achieve more enduring cooperation. As a result, our ability to work together has the potential to rise even further to reach a new pitch of harmony and unity. Again I should stress that I mean more than simply cooperation today and in the here and now. Thanks to the extraordinary mechanism of indirect reciprocity, language can unite the interests of the past, the present, and the future, too.

Many already like to talk of their duty to future generations. Discussions of sustainability focus on the idea of intergenerational equity—providing the next generation, and the ones that follow, with the same environmental potential as presently exists. Politicians often meditate on the legacy that they would like to leave to their grandchildren. I would like to push this idea to its logical conclusion and encourage everyone to think objectively about how they are cooperating with future generations. We need to broaden our horizon of concerns far beyond the events of tomorrow. We need our duty of care to extend to those who have yet to be born. We must do our utmost to cooperate with the many tens of billions of people who will inherit the world from us.

I am hopeful that, in the distant future, a SuperCooperator will gaze toward an infinite blue horizon of opportunities. If we take a cosmic view, there's hope for life. Across the universe I am sure there are thousands of societies, even millions, that are as advanced as our own, if not more so. Each one will no doubt use different approaches to solve the problem of efficient global cooperation. Some will work. Others will fail.

A higher level of selection will operate among them. Some civilizations will expand and prosper in the long term. Many will not and may well perish and wink out. And there are other fates in store. A few will lose their home world only to colonize new planets. Some may die off but leave a new kind of life in their wake, a spacefaring flotilla of intelligent robots that can reproduce, thrive, and explore their galaxy. Those civilizations that have solved the problem of cooperation will persist in the cosmos. We can only hope that this list of successful SuperCooperators will include those carbon-based life forms that we call human beings. In this great adventure, everyone has a role to play. Success depends on all of us. Over to you.

ACKNOWLEDGMENTS

Our efforts to describe the mechanisms of cooperation have depended on the help, kindness, and generosity of a vast number of people. Cooperation rules!

The initial impetus for us to work together came from Roger, when he realized that over two decades he had covered a great many of the papers that I had written for journals such as *Nature* and *Science* for his newspaper, *The Daily Telegraph*. He felt that a book on my work could give a general reader a panoramic view of the latest developments in many scientific disciplines.

The first time I talked about the "five rules for the evolution of cooperation" was at a conference held in March 2006 at the Zoology Department at the University of Oxford in honor of the seventieth birthday of my great friend and mentor, Lord May of Oxford. The warmth and enthusiasm that greeted my lecture was a great reminder of the place where I had been so happy for many years.

Various topics of this book were presented and developed at the Templeton Lectures, which I delivered at Johns Hopkins University in March 2010. These lectures were organized by Steven Gross and funded by a grant from the Metanexus Institute, together with matching funds from the John Templeton Foundation and Johns Hopkins University's Krieger School of Arts and Sciences.

Many thanks to our agent, John Brockman, along with Katinka Matson and Max Brockman. We are also in debt to our editors at Free Press and Canongate, Hilary Redmon and Nick Davies, for their invaluable feedback, advice, and support. Thanks as well to Sydney Tanigawa, Kathryn Higuchi, Amy Ryan, and production colleagues.

Roger was fortunate in that his publisher, David Wilson, and editor in chief, Jeremy Webb, were happy that he pursued this project when he moved in 2008 from *The Daily Telegraph* to become the editor of *New Scientist* magazine.

Roger also benefited from a great deal of support from his friends. Anne Blumberg and Jon Dorfman looked after him during several visits to Cambridge over the years. Valuable opportunities to work on the manuscript were provided by Jean and Barry Blumberg in Rangeley, Maine; Tina and Michael Mahony in Southwold, Suffolk; and Jim Lawrie and Carole Gannon in Fordingbridge, Hampshire. Jamie Lyle and Tracy Nixon provided a lift from Biggin Hill to Bangor, Maine, and then from Rangeley to Bar Harbor and Boston. Several of his friends commented on drafts, notably George Blumberg, Stefano Blumberg, Peter Coveney, Jon Dorfman, Graham Farmelo, Eamonn Matthews, and Brian Millar. Roger's wife, Julia Brookes, has also offered much invaluable advice. And there are others who provided valuable indirect support, such as Paul Carter, Gulshan Chunara, Heather Gething, Raj Persaud, and David Johnson.

Many academics have been kind in answering his questions, such as Anna Dreber Almenberg, Tim Clutton-Brock, Molly Crocket, Ernst Fehr, Erez Lieberman, Bob May, Manfred Milinski, Paul Nurse, David Rand, Alan Sanfey, Hava Siegelmann, Karl Sigmund, Jack Szostak, Corina Tarnita, Manfred Wenzel, and E. O. Wilson. Some have commented on appropriate sections. Oren Harman provided a manuscript of his biography, *The Price of Altruism,* before it was published, along with a draft paper on the origins of George Price's work with John Maynard Smith.

I am grateful to Jeffrey Epstein, the Templeton Foundation, the National Science Foundation, and the National Institutes of Health for their support over the years. I would like to thank Harvard University for allowing me to lead a life that is dedicated to research and teaching in the most exalted and challenging of academic environments. The Program for Evolutionary Dynamics would not exist without the dedicated management of May Huang and the support of Lydia Liu.

Over the years I was fortunate enough to cooperate with a great number of impressive scientists. Here I can only name a few of them, many of whom are my coauthors on various papers that deal with cooperation. I would like to thank: Tibor Antal, Ramy Arnaout, Heather Battey, Nicholas Beale, Niko Beerenwinkel, Baruch Blumberg, Maarten Boerlijst, Immanuel Bomze, Sebastian Bonhoeffer, Pedro Bordalo, Ivana Božic, Hannelore Brandt, Reinhard Bürger, Matteo Cavaliere, Krishnendu Chatterjee, Sarah Coakley, Karen Croxson, Attila Csikasz-Nagy, David Dingli, Anna Dreber Almenberg, Michael Doebeli, Tore Ellingsen, Ernst Fehr, Steven Frank, Feng Fu, Drew Fudenberg, Simon Gaechter, David Haig, Christoph Hauert, Alison Hill, Lorens Imhof, Jeffrey Ishizuka, Yoh Iwasa, Joe Jackson, Vincent Jansen, Timothy Killingback, Natalia Komarova, David Krakauer, Philipp Langer, Simon Levin, Erez Lieberman Aiden (after he got married, he and his wife decided to add an additional name, Aiden, to both their surnames), Michael Manapat, Barnaby Marsh, Robert May, John Maynard Smith, Jean-Baptiste Michel, Franziska Michor, Manfred Milinski, Garrett Mitchener, Charles Nathanson, Partha Niyogi, Hisashi Ohtsuki, Jorge Pacheco, Karen Page, Christina Pawlowitsch, Steven Pinker, Thomas Pfeiffer, Joshua Plotkin, David Rand, Sébastien Roch, Daniel Rosenbloom, Akira Sasaki, Peter Schuster, Anirvan Sengupta, Noam Shoresh, Edward Stabler, Hava Siegelmann, Karl Sigmund, Tina Tang, Corina Tarnita, Christine Taylor, Peter Taylor, David Tilman, Arne Traulsen, Robert Trivers, Matthijs van Veelen, Bert Vogelstein, Neal Wadwha, Nick Wage, Joe Yuichiro Wakano, John Wakeley, Long Wang, Claus Wedekind, Martin Willensdorfer, Edward O. Wilson, and Dominik Wodarz. I am indebted to the French master chef Raymond Ost for exploring culinary forms of cooperation.

Finally, we would both like to thank our families for tolerating our absences while toiling over the manuscript and for giving us support for what turned out to be a much bigger project than either of us realized. We thank Franz, Doris, Betty, Gertrude, Julia, Holly, Philipp, Rory, Sebastian, and Ursula for their patience and love, good humor, and unconditional cooperation.

Martin Nowak and Roger Highfield
Harvard University, January 2011

REFERENCES

AND FURTHER READING

Chapter 0: The Prisoner's Dilemma

References

Ackermann, M., B. Stecher, N. E. Freed, P. Songhet, W. D. Hardt, and M. Doebeli. 2008. Self-destructive cooperation mediated by phenotypic noise. *Nature* 454: 987–90.

Antal, T., M. A. Nowak, and A. Traulsen. 2009. Strategy abundance in 2x2 games for arbitrary mutation rates. *J. Theor. Biol.* 257: 340–44.

Antal, T., A. Traulsen, H. Ohtsuki, C. E. Tarnita, and M. A. Nowak. 2009. Mutation—selection equilibrium in games with multiple strategies. *J. Theor. Biol.* 258: 614–22.

Doebeli, M., and I. Ispolatov. 2010. Complexity and diversity. *Science* 328: 494–97.

Hauert, C., F. Michor, M. A. Nowak, and M. Doebeli. 2006. Synergy and discounting of cooperation in social dilemmas. *J. Theor. Biol.* 239: 195–202.

Helbing, D. 1996. A stochastic behavioral model and a "microscopic" foundation of evolutionary game theory. *Theor. Decis.* 40: 149–79.

Levin, S. A. 2006. Fundamental questions in biology. *PLoS Biol.* 4(9): 1471–72.

May, R. M. 2004. Uses and abuses of mathematics in biology. *Science* 303: 790–93.

Maynard Smith, J., and G. R. Price. 1973. Logic of animal conflict. *Nature* 246: 15–18.

Morris, S. C. 2010. Evolution: Like any other science it is predictable. *Philos. T. Roy. Soc. B* 365: 133–45.

Nash, J. F. 1950. Equilibrium points in n-person games. *Proc. Natl. Acad. Sci. USA* 36: 48–49.

Nowak, M. A. 2004. Prisoners of the dilemma. *Nature* 427: 491.

———. Theory is available light. *Curr. Biol.* 14: R406—R407.

Nowak, M. A., and K. Sigmund. 2004. Evolutionary dynamics of biological games. *Science* 303: 793–99.

Nowak, M. A., A. Sasaki, C. Taylor, and D. Fudenberg. 2004. Emergence of cooperation and evolutionary stability in finite populations. *Nature* 428: 646–50.

Traulsen, A., C. Hauert, H. De Silva, M. A. Nowak, and K. Sigmund. 2009. Exploration dynamics in evolutionary games. *P. Natl. Acad. Sci. USA* 106: 709–12.

Books and Reports

Binmore, K. 1994. *Game theory and the social contract.* Cambridge, MA: MIT Press.

Bürger, R. 2000. *The mathematical theory of selection, recombination, and mutation.* Chichester, UK: Wiley.

Cressman, R. 2003. *Evolutionary dynamics and extensive form games*. Cambridge, MA: MIT Press.

Ewens, W. J. 2004. *Mathematical population genetics*. 2nd ed. Berlin: Springer.

Fisher, R. A. 1930. *The genetical theory of natural selection*. Oxford: Oxford University Press.

Fudenberg, D., and K. Levine. 1998. *The theory of learning in games*. Cambridge, MA: MIT Press.

Fudenberg, D., and J. Tirole. 1991. *Game theory*. Cambridge, MA: MIT Press.

Gintis, H. 2009. *Game theory evolving*. Princeton: Princeton University Press.

Goethe, J. W. von. 1996. *Faust*. Trans. W. Arndt. New York: W.W. Norton.

Haldane, J. B. S. 1932. *The causes of evolution*. London: Longmans, Green.

Hamilton, W. D. 1996. *Evolution of social behaviour*. Vol. 1 of *Narrow roads of gene land*. New York: W. H. Freeman.

———.2001. *Evolution of sex*. Vol. 2 of *Narrow roads of gene land*. Oxford: Oxford University Press.

Jones, S. 2009. *Darwin's island: The Galapagos in the garden of England*. London: Little Brown.

Judson, H. 1996. *The eighth day of creation: Makers of the revolution in biology*. Cold Spring Harbor, NY: Cold Spring Harbor Laboratory Press.

Kimura, M. 1983. *The neutral theory of molecular evolution*. Cambridge: Cambridge University Press.

May, R. M. 1973. *Stability and complexity in model ecosystems*. Princeton: Princeton University Press.

Maynard Smith, J. 1982. *Evolution and the theory of games*. Cambridge: Cambridge University Press.

Mayr, E. 2001. *What evolution is*. New York: Basic Books.

Moran, P. A. P. 1962. *The statistical processes of evolutionary theory*. Oxford: Clarendon Press.

Nowak, M. A. 2006. *Evolutionary dynamics: Exploring the equations of life*. Cambridge, MA: Belknap Press of Harvard University Press.

Poundstone, W. 1993. *Prisoner's dilemma*. London: Anchor.

Rapoport, A., and A. M. Chammah. 1965. *Prisoner's dilemma*. Ann Arbor, MI: University of Michigan Press.

Sigmund, K. 1993. *Games of life: Explorations in ecology, evolution and behaviour*. Oxford: Oxford University Press.

———. 2010. *The calculus of selfishness*. Princeton: Princeton University Press.

von Neumann, J., and O. Morgenstern. 1944. *Theory of games and economic behavior*. Princeton: Princeton University Press.

Weibull, J. W. 1995. *Evolutionary game theory*. Cambridge, MA: MIT Press.

Chapter 1: Direct Reciprocity—Tit for Tat

References

Axelrod, R., and D. Dion. 1988. The further evolution of cooperation. *Science*: 242: 1385–90.

Axelrod, R., and W. D. Hamilton. 1981. The evolution of cooperation. *Science* 211: 1390–96.

Boerlijst, M. C., M. A. Nowak, and K. Sigmund. 1997. The logic of contrition. *J. Theor. Biol.* 185: 281–93.

Clutton-Brock, T. 2009. Cooperation between non-kin in animal societies. *Nature* 462: 51–57.

de Waal, F. B. M. 1997. The chimpanzee's service economy: Food for grooming. *Evol. Hum. Behav.* 18: 375–86.

Fudenberg, D., and E. Maskin. 1986. The folk theorem in repeated games with discounting or with incomplete information. *Econometrica* 50: 533–54.

———. 1990. Evolution and cooperation in noisy repeated games. *Am. Econ. Rev.* 80: 274–79.

Gumert, M. D. 2007. Payment for sex in a macaque mating market. *Anim. Behav.* 74: 1655–67.

Hart, B. L., and L. A. Hart. 1992. Reciprocal allogrooming in impala, *Aepyceros melampus*. *Anim. Behav.* 44: 1073–83.

Imhof, L. A., D. Fudenberg, and M. A. Nowak. 2005. Evolutionary cycles of cooperation and defection. *Proc. Natl. Acad. Sci. USA* 102: 10797–800.

Imhof, L. A., and M. A. Nowak. 2010. Stochastic evolutionary dynamics of direct reciprocity. *P. Roy. Soc. Lond. B Bio.* 277: 463–68.

Kraines, D., and V. Kraines. 1989. Pavlov and the prisoner's dilemma. *Theor. Decis.* 26: 47–79.

May, R. M. 1976. Simple mathematical models with very complicated dynamics. *Nature* 261: 459–67.

———.1987. More evolution of cooperation. *Nature* 327: 15–17.

Melis, A. P., B. Hare, and M. Tomasello. 2008. Do chimpanzees reciprocate received favours? *Anim. Behav.* 76: 951–62.

Milinski, M. 1987. Tit for tat in sticklebacks and the evolution of cooperation. *Nature* 325: 433–35.

Mitani, J. C., and D. P. Watts. 2001. Why do chimpanzees hunt and share meat? *Anim. Behav.* 61: 915–24.

Nowak, M. A. 2004. Prisoners of the dilemma. *Nature* 427: 491.

Nowak, M. A., R. M. Anderson, M. C. Boerlijst, S. Bonhoeffer, R. M. May, and A. J. McMichael. 1996. HIV-1 evolution and disease progression. *Science* 274: 1008–1011.

Nowak, M. A., R. M. Anderson, A. R. McLean, T. Wolfs, J. Goudsmit, and R. M. 1991. Antigenic diversity thresholds and the development of AIDS. *Science* 254 (May): 963–69.

Nowak, M. A., R. M. May, R. E. Phillips, S. Rowland-Jones, D. G. Lalloo, S. McAdam, P. Klenerman, B. Köppe, K. Sigmund, C. R. M. Bangham, and A. J. McMichael. 1995. Antigenic oscillations and shifting immunodominance in HIV-1 infections. *Nature* 375: 606–11.

Nowak, M. A., R. M. May, and K. Sigmund. 1995. The arithmetics of mutual help. *Sci. Am.* 272: 76–81.

Nowak, M. A., and A. J. McMichael. 1995. How HIV defeats the immune system. *Sci. Am.* 273: 58–65.

Nowak, M. A., A. Sasaki, C. Taylor, and D. Fudenberg. 2004. Emergence of cooperation and evolutionary stability in finite populations. *Nature* 428: 646–50.

Nowak, M. A., and K. Sigmund. 1989. Oscillations in the evolution of reciprocity. *J. Theor. Biol.* 137: 21–26.

————.1992. Tit for tat in heterogeneous populations. *Nature* 355: 250–53.

————.1993. A strategy of win-stay, lose-shift that outperforms tit-for-tat in the prisoner's dilemma game. *Nature* 364: 56–58.

————.1994. The alternating prisoner's dilemma. *J. Theor. Biol.* 168: 219–26.

Ogg, G. S., X. Jin, S. Bonhoeffer, P. R. Dunbar, M. A. Nowak, S. Monard, J. P. Segal, Y. Cao, S. L. Rowland-Jones, V. Cerundolo, A. Hurley, M. Markowitz, D. D. Ho, D. F. Nixon, and A. J. McMichael. 1998. Quantitation of HIV-1 specific cytotoxic T lymphocytes and plasma load of viral RNA. *Science* 279: 2103–6.

Oskamp, S. 1971. Effects of programmed strategies on cooperation in the prisoner's dilemma and other mixed-motive games. *J. Conflict Resolution* 15: 225–59.

Packer, C. 1977. Reciprocal altruism in olive baboons (*Papio anubis*). *Nature* 265: 441–43.

Ratnayeke, S. M., and W. P. J. Dittus. 1989. Individual and social behavioral responses to injury in wild toque macaques (*Macaca sinica*). *Int. J. Primatol.* 10: 215–33.

Riskin, D., and J. W. Hermanson. 2005. Biomechanics: Independent evolution of running in vampire bats. *Nature* 434: 292.

Trivers, R. L. 1971. The evolution of reciprocal altruism. *Q. Rev. Biol.* 46: 35–57.

Wahl, L. M., M. A. Nowak. 1999. The continuous prisoner's dilemma: I. Linear reactive strategies. *J. Theor. Biol.* 200: 307–21.

Wedekind, C., and M. Milinski. 1996. Human cooperation in the simultaneous and the alternating prisoner's dilemma: Pavlov versus generous tit-for-tat. *P. Natl. Acad. Sci. USA* 93: 2686–89.

Wiegrebe, L., and U. Groeger. 2006. Classification of human breathing sounds by the common vampire bat, *Desmodus rotundus*. *BMC Biol.* 4:18.

Wilkinson, G. S. 1984. Reciprocal food sharing in the vampire bats. *Nature* 308: 181–84.

Wilson, W. 1971. Reciprocation and other techniques for inducing cooperation in the prisoner's dilemma game. *J. Conflict Resolution* 15: 167–95.

Books and Reports

Anderson, R. M., and R. M. May. 1991. *Infectious diseases of humans.* Oxford: Oxford University Press.

Axelrod, R. 1984. *The evolution of cooperation.* London: Penguin.

————.1997. *The complexity of cooperation: Agent-based models of competition and collaboration.* Princeton: Princeton University Press.

Dugatkin, L. A. 1997. *Cooperation among animals.* Oxford: Oxford University Press.

Hammerstein, P., ed. 2003. *Genetic and cultural evolution of cooperation.* Cambridge, MA: MIT Press.

Hofbauer, J.,and K. Sigmund. 1998. *Evolutionary games and population dynamics.* Cambridge: Cambridge University Press.

May, R. M. 2001. *Stability and complexity in model ecosystems.* Princeton: Princeton University Press.

Maynard Smith, J. 1982. *Evolution and the theory of games.* Cambridge: Cambridge University Press.

Nowak, M. A., and R. M. May. 2000. *Virus dynamics: Mathematical principles of immunology and virology.* Oxford: Oxford University Press.

Ridley, M. 1996. *The origins of virtue.* London: Viking.

Sigmund, K. 1996. *Games of life: Explorations in ecology, evolution, and behaviour*. Oxford: Oxford University Press.

Weibull, J. 1995. *Evolutionary game theory.* Cambridge, MA: MIT Press.

Websites

Internet Classics Archive. *Nicomachean ethics*. Aristotle. Trans. W. D. Ross.
   http://classics.mit.edu/Aristotle/nicomachaen.5.v.html

Internet Classics Archive. *Crito*. Plato. Trans Benjamin Jowett.
   http://classics.mit.edu/Plato/crito.html

Nobel Prize website. Robert Aumann, autobiography and further information.
   http://nobelprize.org/nobel_prizes/economics/laureates/2005/aumann-autobio.html
   http://nobelprize.org/nobel_prizes/economics/laureates/2005/ecoadv05.pdf

Emails

Craig Packer, email message to Roger Highfield, January 1, 2010

Chapter 2: Indirect Reciprocity—The Power of Reputaton

References

Brandt, H., and K. Sigmund. 2005. Indirect reciprocity, image scoring and moral hazard. *Proc. Natl. Acad. Sci. USA* 102: 2666–70.

Ehrlich, P. R., and S. A. Levin. 2005. The evolution of norms. *PLoS Biol.* 3(6): 943–48.

Henrich, J., and R. Boyd. 2008. Division of labor, economic specialization and the evolution of social stratification. *Curr. Anthropol.* 49: 715–24.

Izuma, K., D. N. Saito, and N. Sadato. 2008. Processing of social and monetary rewards in the human striatum. *Neuron* 58: 284–94.

Kandori, M. 1992. Social norms and community enforcement. *Rev. Econ. Stud.* 59: 63–80.

Leimar, O., and P. Hammerstein. 2001. Evolution of cooperation through indirect reciprocation. *P. Roy. Soc. Lond. B Bio* 268: 745–53.

Nowak, M. A., and K. Sigmund. 1998. Evolution of indirect reciprocity by image scoring. *Nature* 393: 573–77.

———.2005. Evolution of indirect reciprocity. *Nature* 437: 1291–98.

Ohtsuki, H., and Y. Iwasa. 2004. How should we define goodness? Reputation dynamics in indirect reciprocity. *J. Theor. Biol.* 231: 107–120.

———.2006. The leading eight: Social norms that can maintain cooperation by indirect reciprocity." *J. Theor. Biol.* 239: 435–44.

Ohtsuki, H., Y. Iwasa, and M. A. Nowak. 2009. Indirect reciprocity provides only a narrow margin of efficiency for costly punishment. *Nature* 457: 79–82.

Panchanathan, K., and R. Boyd. 2004. Indirect reciprocity can stabilize cooperation without the second-order free-rider problem. *Nature* 432: 499–502.

Rutte, C., and M. Taborsky. 2007. Generalized reciprocity in rats. *PLoS Biol.* 5: e196.

Wedekind, C., and M. Milinski. 2000. Cooperation through image scoring in humans. *Science* 288: 850–52.

Books and Reports

Alexander, R. D. 1987. *The biology of moral systems*. New York: Aldine de Gruyter.

Deutsch, O. E. 1958. *Schubert: Memoirs by his friends*. London: Dent.

Gibbs, C. H. 2000. *The life of Schubert*. Cambridge: Cambridge University Press.

Sober, E., and D. S. Wilson. 1998. *Unto others: The evolution and psychology of unselfish behavior*. Cambridge, MA: Harvard University Press.

Sugden, R. 1986. *The economics of rights, cooperation and welfare*. Oxford, NY: Blackwell.

Websites

The Project Gutenberg EBook of Cyropaedia, by Xenophon. Ed. F. M. Stawell Trans. Henry Graham Dakyns, www.gutenberg.org/dirs/2/0/8/2085/2085.txt

Chapter 3: Spacial Games—Chessboard of Life

References

Durrett, R., and S. A. Levin. 1994. The importance of being discrete (and spatial). *Theor. Popul. Biol.* 46: 363–94.

———. 1994. Stochastic spatial models: A user's guide to ecological applications. *Philos. T. Roy. Soc. B* 343: 329–50.

Ellison, G. 1993. Learning, local interaction, and coordination. *Econometrica* 61: 1047–71.

Fu, F., M. A. Nowak, and C. Hauert. 2010. Invasion and expansion of cooperators in lattice populations: Prisoner's dilemma vs. Snowdrift games. *J. Theor. Biol.* 266: 358–66.

Hassell, M. P., H. N. Comins, and R. M. May. 1991. Spatial structure and chaos in insect population dynamics. *Nature* 353: 255–58.

———. 1994. Species coexistence and self-organizing spatial dynamics. *Nature* 370: 290–92.

Hauert, C. 2002. Effects of space in 2×2 games. *Int. J. Bifurcat. Chaos* 12: 1531–48.

Hauert, C., and M. Doebeli. 2004. Spatial structure often inhibits the evolution of cooperation in the snowdrift game. *Nature* 428: 643–46.

Helbing, D., and W. Yu. 2009. The outbreak of cooperation among success-driven individuals under noisy conditions. *P. Natl. Acad. Sci. USA* 106: 3680–85.

Killingback, T., and M. Doebeli. 1998. Self-organized criticality in spatial evolutionary game theory. *J. Theor. Biol.* 191: 335–40.

Langer P., M. A. Nowak, and C. Hauert. 2008. Spatial invasion of cooperation. *J. Theor. Biol.* 250: 634–41.

Levin, S. A., and R. T. Paine. 1974. Disturbance, patch formation, and community structure. *P. Natl. Acad. Sci. USA* 71: 2744–47.

Nakamaru, M., H. Matsuda, and Y. Iwasa. 1997. The evolution of cooperation in a lattice structured population. *J. Theor. Biol.* 184: 65–81.

Nowak, M. A., and R. May. 1992. Evolutionary games and spatial chaos. *Nature* 359: 826–29.

———. 1993. The spatial dilemmas of evolution. *Int. J. Bifurcat. Chaos* 3: 35–78.

Nowak, M. A., S. Bonhoeffer, and R. M. May. 1994. More spatial games. *Int. J. Bifurcat. Chaos* 4: 33–56.

Tilman, D., R. M. May, C. L. Lehman, and M. A. Nowak. 1994. Habitat destruction and the extinction debt. *Nature* 371: 65–66.

Books and Reports

Brand, S. 2009. *Whole earth discipline*. London: Atlantic Books.

Cairns-Smith, A. G. 1982. *Genetic takeover and the mineral origins of life*. Cambridge: Cambridge University Press.

Highfield, R., and P. Coveney. 1992. *The arrow of time: The quest to solve science's greatest mystery*. New York: Ballantine Books.

———. 1996. *Frontiers of complexity: The search for order in a chaotic world*. London: Faber and Faber.

Holland, J. 1999. *Emergence: From chaos to order*. New York: Basic Books.

Nowak, M. A. 2006. *Evolutionary dynamics: Exploring the equations of life*. Cambridge, MA: Harvard University Press.

Renfrew, A. C. 2003. *Figuring it out: The parallel visions of artists and archaeologists*. London: Thames and Hudson.

Tilman, D., and P. Karieva, eds. 1997. *Spatial ecology: The role of space in population dynamics and interspecific interactions*. Princeton: Princeton University Press.

von Neumann, J. 1966. *Theory of self-reproducing automata*. Champaign: University of Illinois Press.

Wolfram, S. 2002. *A new kind of science*. Champaign, IL: Wolfram Media.

Articles

Highfield, R. 2003. The birth of our modern minds. *The Daily Telegraph,* October 15, Science Page.

Websites

Christoph Hauert's VirtualLabs:
  http://www.math.ubc.ca/~hauert/

Interviews

Robert May. 2008. Interview by Roger Highfield. September 15.

Chapter 4: Group Selection—Tribal Wars

References

Bijma, P., and M. J. Wade. 2008. The joint effects of kin, multilevel selection and indirect genetic effects on response to genetic selection. *J. Evol. Biol.* 21: 1175–88.

Bowles, S. 2006. Group competition, reproductive leveling, and the evolution of human altruism. *Science* 314: 1569–72.

Boyd, R., and P. J. Richerson. 1990. Group selection among alternative evolutionarily stable strategies" *J. Theor. Biol.* 145: 331–42.

Craig, D. M. 1982. Group selection versus individual selection: An experimental analysis. *Evolution* 36: 271–82.

Fletcher, J. A., and M. Zwick. 2004. Strong altruism can evolve in randomly formed groups. *J. Theor. Biol.* 228: 303–13.

Kerr, B., C. Neuhauser, B. J. Bohannan, and A. M. Dean. 2006. Local migration promotes competitive restraint in a host pathogen tragedy of the commons. *Nature* 442: 75–78.

Killingback, T., J. Bieri, and T. Flatt. 2006. Evolution in group-structured populations can resolve the tragedy of the commons. *P. Roy. Soc. Lond. B Bio.* 273: 1477–81.

Newton, I. 1999. In Memoriam: V. C. Wynne-Edwards. *The Auk* 116: 815–16.

Rogers, D. S., and P. R. Ehrlich. 2008. Natural selection and cultural rates of change. *P. Natl. Acad. Sci. USA* 105: 3416–20.

Rogers, D. S., M. W. Feldman, and P. R. Ehrlich. 2007. Inferring population histories using cultural data. *P. Roy. Soc. Lond. B Bio.* 276: 3835–43.

Szathmary, E., and L. Demeter. 1987. Group selection of early replicators and the origin of life. *J. Theor. Biol.* 128: 463–86.

Traulsen, A., A. M. Sengupta, and M. A. Nowak. 2005. Stochastic evolutionary dynamics on two levels. *J. Theor. Biol.* 235: 393–401.

Traulsen A., and Martin Nowak. 2006. Evolution of cooperation by multilevel selection. *P. Natl. Acad. Sci. USA* 103: 10952–55.

Traulsen A., N. Shoresh, and M. A. Nowak. 2008. Analytical results for individual and group selection of any intensity. *Bull. Math. Biol.* 70: 1410–24.

Wade, M. J. 1976. Group selections among laboratory populations of tribolium. *P. Natl. Acad. Sci. USA* 73: 4604–07.

Wilson, D. S. 1975. A theory of group selection. *P. Natl. Acad. Sci. USA* 72: 143–46.

Wilson, D. S., and E. O. Wilson. 2007. Rethinking the theoretical foundation of sociobiology. *Q. Rev. Biol.* 82: 327–48.

Books and Reports

Darwin, C. 1871. *The descent of man, and selection in relation to sex.* London: John Murray.

Dawkins, R. 2009. *The greatest show on earth: The evidence for evolution.* London: Bantam Press.

Hamilton, W. D. 2006. *Narrow roads of gene land 1: Evolution of social behavior.* Oxford: W. H. Freeman.

Sober, E., and D. S. Wilson. 1998. *Unto others: The evolution and psychology of unselfish behavior.* Cambridge, MA: Harvard University Press.

Wynne-Edwards, V. C. 1962. *Animal dispersion in relation to social behavior.* New York: Hafner.

Websites

Racey, P. A. Royal Society of Edinburgh. *Obituary of Vero Wynne-Edwards.* www.royalsoced.org.uk/fellowship/obits/obits_alpha/wynne-edwards_vero.pdf

Interviews

Jack Szostak. 2010. Interview by Roger Highfield. April.

Chapter 5: Kin Selection—Nepotism

References

Cavalli-Sforza, L. L., and M. W. Feldman. 1978. Darwinian selection and "altruism." *Theor. Popul. Biol.* 14: 268–80.

Doebeli, M., and C. Hauert. 2006. Limits to Hamilton's rule. *J. Evol. Biol.* 19: 1386–88.

Fletcher, J. A., and M. Doebeli. 2009. A simple and general explanation for the evolution of altruism. *P. Roy. Soc. Lond. B Bio.* 276: 13–19.

Fletcher, J. A., M. Zwick, M. Doebeli, and D. S. Wilson. 2006. What's wrong with inclusive fitness? *Trends Ecol. Evol.* 21: 597–98.

Foster, K. R., T. Wenseleers, and F. L.W. Ratnieks. 2006. Kin selection is the key to altruism. *Trends Ecol. Evol.* 21: 57–60.

Frank, S. A. 1995. George Price's contributions to evolutionary genetics. *J. Theor. Biol.* 175: 373–388.

Grafen, A. 1984. Natural selection, kin selection and group selection. In J. R. Krebs and N. B. Davies, eds., *Behavioural Ecology,* 62–84. Oxford: Blackwell Scientific Publications.

Grafen, A. 2004. William Donald Hamilton. *Biogr. Mems Fell. R. Soc. Lond.* 50: 109–32.

Haldane, J. B. S. 1955. Population Genetics. *New Biology* 18: 34–51.

Hamilton, W. D. 1964. The genetical evolution of social behaviour I and II. *J. Theor. Biol.* 7: 1–16 and 17–52.

Hamilton, W. D. 1991. My intended burial and why. *Insectarium* 28: 238–47.

Karlin, S., and C. Matessi. 1983. Kin selection and altruism. *P. Roy. Soc. Lond. B Bio.* 219: 327–53.

Lehmann, L., and L. Keller. 2006. The evolution of cooperation and altruism—a general framework and a classification of models. *J. Evol. Biol.* 19: 1365–76.

Maynard Smith, J. 1965. Obituary of Haldane. *Nature* 4981: 239.

Maynard Smith, J., and G. R. Price. 1972. The logic of animal conflict. *Nature* 246: 15–18.

Mehdiabadi, N. J., H. K. Reeve, and U. G. Mueller. 2003. Queens versus workers: Sex-ratio conflict in eusocial Hymenoptera. *Trends Ecol. Evol.* 18: 88–93.

Michod, R. E., and W. D. Hamilton. 1980. Coefficients of relatedness in sociobiology. *Nature* 288: 694–97.

Nowak, M. A. 2006. Five rules for the evolution of cooperation. *Science* 314: 1560–63.

Nowak, M. A., C. Tarnita, and E. O. Wilson. 2010. The evolution of eusociality. *Nature* 466: 1057–62.

Queller, D. C. 1985. Kinship, reciprocity and synergism in the evolution of social behaviour. *Nature* 318: 366–67.

Queller, D. C., and J. E. Strassmann. 1998. Kin selection and social insects. *Bioscience* 48: 165–75.

Rousset, F., and S. Billiard. 2000. A theoretical basis for measures of kin selection in subdivided populations: Finite populations and localized dispersal. *J. Evol. Biol.* 13: 814–26.

Taylor, P. D. 1992. Altruism in viscous populations—an inclusive fitness model. *Evol. Ecol.* 6: 352–56.

Taylor, P. D., and S. Frank. 1996. How to make a kin selection argument. *J. Theor. Biol.* 180: 27–37.

Traulsen, A. 2010. Mathematics of kin- and group-selection: Formally equivalent? *Evolution* 64: 316–23.

van Veelen M. 2005. On the use of the Price equation. *J. Theor. Biol.* 237: 412–26.

van Veelen, M. 2009. Group selection, kin selection, altruism and cooperation: When inclusive fitness is right and when it can be wrong. *J. Theor. Biol.* 259: 589–600.

Wade, M. J., D. S. Wilson, C. Goodnight, D. Taylor, Y. Bar-Yam, M. A. M. de Aguiar, B.

Stacey, J. Werfel, G. A. Hoelzer, E. D. Brodie III, P. Fields, F. Breden, T. A. Links-vayer, J. A. Fletcher, P. J. Richerson, J. D. Bever, J. D. Van Dyken, and P. Zee. 2010. Multilevel and kin selection in a connected world. *Nature* 463: E8—E9.

Weiner, J. 2000. A conversation with John Maynard Smith. *Natural History* 109: 78–91.

West, S. A., A. S. Griffin, and A. Gardner. 2007. Evolutionary explanations for coopera-tion. *Curr. Biol.* 17: R661—R672.

West, S. A., A. S. Griffin, and A. Gardner. 2008. Social semantics: How useful has group selection been? *J. Evol. Biol.* 21: 374–85.

Books and Reports

Frank, S. A. 1998. *Foundations of social evolution.* Princeton: Princeton University Press.

Haldane, J. B. S. 1932. *The causes of evolution.* London: Longmans, Green.

Hamilton, W. D. 2006. *Evolution of social behavior.* Vol. 1 of *Narrow roads of gene land.* Oxford: W. H. Freeman.

Harman, O. 2010. *The price of altruism. George Price and the search for the origins of kind-ness.* London: The Bodley Head.

Hunt, J. H. 2007. *The evolution of social wasps.* Oxford: Oxford University Press.

Rousset, F. 2004. *Genetic structure and selection in subdivided populations.* Princeton: Princeton University Press.

Wilson, E. O. 1975. *Sociobiology: The new synthesis.* Cambridge, MA: Harvard University Press.

———. 2006. *Naturalist.* Washington, DC: Island Press.

Articles

Dawkins, R. 2000. W. D. Hamilton (1936–2000). *The Independent.* March 10.

Haldane's quip in a pub can be found in "Accidental Career." 1974. *New Scientist.* August 8.

Jones, S. 2010. Between kindness and madness. *New Scientist:* May 29.

Maynard Smith, J. 1992. Past Master of the Possible World. *The Guardian.* March 27.

Emails

Oren Harman. 2010. Private communication with Roger Highfield, May 31.

Chapter 6: Prelife

References

Bartel, D. P., and J. W. Szostak.1993. Isolation of new ribozymes from a large pool of ran-dom sequences. *Science* 261: 1411–18.

Benner, S. A., M. D. Caraco, J. M. Thomson, and E. A. Gaucher. 2002. Planetary biology—paleontological, geological, and molecular histories of life. *Science* 296: 864–68.

Cech, T. R. 1993. The efficiency and versatility of catalytic RNA: Implications for an RNA world. *Gene* 135: 33–36.

Chen, I. A., R. W. Roberts, and J. W. Szostak. 2004. The emergence of competition between model protocells. *Science* 305: 1474–76.

Crick, F. H. 1968. The origin of the genetic code. *J. Mol. Biol.* 38: 367–79.

Crow, James. 1988. Sewall Wright, 1889–1988. *Jpn. J. Genet.* 63: 217–18.

Eigen, M. 1971. Molecular self-organization and the early stages of evolution. *Q. Rev. Biophys.* 4: 149–212.

Eigen, M., J. McCaskill, and P. Schuster. 1989. The molecular quasi-species. *Adv. Chem. Phys.* 75: 149–263.

Eigen, M., and P. Schuster. 1977. The hyper cycle: A principle of natural self-organization. Part A: Emergence of the hyper cycle. *Naturwissenschaften* 64: 541–65.

Fontana, W., and L. W. Buss. 1994. The arrival of the fittest: Toward a theory of biological organization. *B. Math. Biol.* 56: 1–64.

Joyce, G. F. 1989. RNA evolution and the origins of life. *Nature* 338: 217–24.

———. 2002. The antiquity of RNA-based evolution. *Nature* 418: 214–21.

Manapat, M. L., I. A. Chen, and M. A. Nowak. 2010. The basic reproductive ratio of life. *J. Theor. Biol.* 263: 317–27.

Manapat, M., H. Ohtsuki, R. Bürger, and M. A. Nowak (2009). Originator dynamics. *J. Theor. Biol.* 256: 586–95.

Nowak, M. A. 1992. What is a quasispecies? *Trends Ecol. Evol.* 7: 118–21.

Nowak, M. A., M. C. Boerlijst, J. Cooke, and J. Maynard Smith. 1997. Evolution of genetic redundancy. *Nature* 388: 167–71.

Nowak, M. A., and H. Ohtsuki. 2008. Prevolutionary dynamics and the origin of evolution. *P. Natl. Acad. Sci. USA* 105: 14924–27.

Nowak, M., and P. Schuster. 1989. Error thresholds of replication in finite populations mutation frequencies and the onset of Muller's ratchet. *J. Theor. Biol.* 137: 375–95.

Ohtsuki, H., and M. A. Nowak. 2009. Prelife catalysts and replicators. *P. Roy. Soc. Lond. B Bio.* 276: 3783–90.

Rajamani, S., J. K. Ichida, T. Antal, D. A. Treco, K. Leu, M. A. Nowak, J. W. Szostak, and I. A. Chen. 2010. Effect of stalling after mismatches on the error catastrophe in nonenzymatic nucleic acid replication. *J. Am. Chem. Soc.* 132: 5880–85.

Santos, M. 1998. Origin of chromosomes in response to mutation pressure. *Am. Nat.* 152: 751–56.

Sievers, D., and G. von Kiedrowski. 1994. Self-replication of complementary nucleotide-based oligomers. *Nature.* 369: 221–24.

Szostak, J. W., D. P. Bartel, and P. L. Luisi. 2001. Synthesizing life. *Nature* 409: 387–90.

Weiner, J. 2000. A Conversation with John Maynard Smith. *Nat. Hist.* 109: 78–91.

Books and Reports

Eigen, M. 1992. *Steps towards life: A perspective on evolution.* Oxford: Oxford University Press.

Brand, Stewart. 2009. *Whole earth discipline.* London: Atlantic Books.

Kauffman, S. A. 1993. *The origins of order: Self-organization and selection in evolution.* New York: Oxford University Press.

Maynard Smith, J., and E. Szathmary. 1999. *The origins of life.* Oxford: Oxford University Press.

Chapter 7: Society of Cells

References

Abbott, L. H., and F. Michor. 2006. Mathematical models of targeted cancer therapy. *Brit. J. Cancer* 95: 1136–41.

Allwood, A. C., M. R. Walter, B. S. Kamber, C. P. Marshall, and I. W. Burch. 2006. Stromatolite reef from the Early Archaean era of Australia. *Nature* 441: 714–18.

Cavalier-Smith, T. 2010. Deep phylogeny, ancestral groups and the four ages of life. *Philos. T. Roy. Soc. B* 365:111–32.

Dingli, D., F. Michor, T. Antal, and J. M. Pacheco. 2007. The emergence of tumor metastases. *Cancer Biol. Ther.* 6: 383–90.

Dingli, D., and M. A. Nowak. 2006. Infectious tumour cells. *Nature* 443: 35–36.

Dingli, D., A. Traulsen, and F. Michor. 2007. (A)Symmetric stem cell replication and cancer. *PLoS Comput. Biol.* 3: e53.

Durrett, R., J. Foo, K. Leder, J. Mayberry, and F. Michor. 2010. Evolutionary dynamics of tumor progression with random fitness values. *Theor. Popul. Biol.* 78: 54–66.

Foo, J., and F. Michor. 2010. Evolution of resistance to anti-cancer therapy during general dosing schedules. *J. Theor. Biol.* 263: 179–88.

Frank, S. A., and M. A. Nowak. 2003. Developmental predisposition to cancer. *Nature* 422: 494.

Haeno, H., R. L. Levine, D. G. Gilliland, and F. Michor. 2009. The cell of origin of hematopoietic malignancies. *P. Natl. Acad. Sci. USA* 106: 16616–21.

Haeno, H., and F. Michor. 2010. The evolution of tumor metastases during clonal expansion. *J. Theor. Biol.* 263: 30–44.

Jones, S., W. Chen, G. Parmigiani, F. Diehl, N. Beerenwinkel, T. Antal, A. Traulsen, M. A. Nowak, C. Siegel, V. E. Velculescu, K. W. Kinzler, B. Vogelstein, J. Willis, and S. D. Markowitz. 2008. Comparative lesion sequencing provides insights into tumor evolution. *P. Natl. Acad. Sci. USA* 105: 4283–88.

Koshland, D. E. 1993. Molecule of the year. *Science* 262: 1953.

Komarova, N. L., D. Wodarz. 2010. ODE models for oncolytic virus dynamics. *J. Theor. Biol.* 263: 530–43.

Michor, F. 2008. Mathematical models of cancer stem cells. *J. Clin. Oncol.* 26: 2854–61.

Michor, F., T. P. Hughes, Y. Iwasa, S. Branford, N. P. Shah, C. L. Sawyers, and M. A. Nowak. 2005. Dynamics of chronic myeloid leukemia. *Nature* 435: 1267–70.

Michor, F., Y. Iwasa, and M. A. Nowak. 2004. Dynamics of cancer progression. *Nat. Rev. Cancer* 4: 197–205.

Nadell, C. D., B. L. Bassler, and S. A. Levin. 2008. Minireview: Observing bacteria through the lens of social evolution. *J. Biol.* 7: 27.

Nowak, M. A., N. L. Komarova, A. Sengupta, P. F. Jallepalli, I. M. Shih, B. Vogelstein, and C. Lengauer. 2002. The role of chromosomal instability in tumor initiation. *P. Natl. Acad. Sci. USA* 99: 16226–31.

Nowak, M. A., F. Michor, and Y. Iwasa. 2003. The linear process of somatic evolution. *P. Natl. Acad. Sci. USA* 100: 14966–69.

———. 2004. Evolutionary dynamics of tumor suppressor gene inactivation. *P. Natl. Acad. Sci. USA* 101: 10635–38.

Okamoto, N., and I. Inouye. 2005. A Secondary Symbiosis in Progress? *Science* 310: 287.

Olson, J. M. 2006. Photosynthesis in the Archean era. *Photosyn. Res.* 88: 109–17.

Pepper, J. W., K. Sprouffske, and C. C. Maley. 2007. Animal cell differentiation patterns suppress somatic evolution. *PLoS Comput. Biol.* 3: e250.

Rajagopalan, H., M. A. Nowak, B. Vogelstein, and C. Lengauer. 2003. The significance of unstable chromosomes in colorectal cancer. *Nat. Rev. Cancer* 3: 695–701.

Vogelstein, B., D. Lane, and A. J. Levine. 2000. Surfing the p53 Network. *Nature* 408: 307–10.

Yachida, S., S. Jones, I. Bozic, T. Antal, R. Leary, B. Fu, M. Kamiyama, R H. Hruban, J. R. Eshleman, M. A. Nowak, V. E. Velculescu, K. W. Kinzler, B. Vogelstein, C. A. Iacobuzio-Donahue. 2010. Distant metastasis occurs late during the genetic evolution of pancreatic cancer. *Nature* 467:1114-17.

Books

Miller, S. L., and L. E. Orgel. 1974. *The origins of life on the earth.* Englewood Cliffs, NJ: Prentice-Hall.
Vogelstein, B., and K. W. Kinzler, eds. 1998. *The genetic basis of human cancer.* Toronto: McGraw-Hill.
Wodarz, D., and N. L. Komarova. 2005. *Computational biology of cancer: lecture notes and mathematical modeling.* World Scientific Publishing.
World Health Organisation. 2009. *World health statistics 2009.* Geneva: World Health Organization.

Websites

McNamara, Ken. Stromatolites—great survivors under threat. The Geological Society.
    www.geolsoc.org.uk/gsl/site/GSL/lang/en/pid/6727
Vogelstein, Bert. Academy of Achievement.
    http://www.achievement.org/autodoc/page/vog0int-4
World Health Organisation. Factsheet on Cancer.
    http://www.who.int/mediacentre/factsheets/fs297/en/index.html

Chapter 8: The Lord of the Ants

References

Abouheif, E., and G. A. Wray. 2002. Evolution of the gene network underlying wing polyphenism in ants. *Science* 297: 249–52.
Amdam, G. V., and S. C. Seehuu. 2006. Order, disorder, death: Lessons from a superorganism. *Adv. Cancer Res.* 95: 31–60.
Anderson, M. 1984. The evolution of eusociality. *Annu. Rev. Ecol. Syst.* 15: 165–89.
Boomsma, J. J. 2009. Lifetime monogamy and the evolution of eusociality. *Philos. T. Roy. Soc. B* 364: 3191–207.
Charnov, E. L. 1978. Evolution of eusocial behavior: Offspring choice or parental parasitism? *J. Theor. Biol.* 75: 451–65.
Craig, R. 1979. Parental manipulation, kin selection, and the evolution of altruism. *Evolution* 33: 319–34.
———. 1979. Subfertility and the evolution of eusociality by kin selection. *J. Theor. Biol.* 100: 379–97.
Gadagkar, R. 1990. Origin and evolution of eusociality: A perspective from studying primitively eusocial wasps. *J. Genet.* 69: 113–25.
Hunt, J. H., and G. V. Amdam. 2005. Bivoltinism as an antecedent to eusociality in the paper wasp genus *Polistes.* *Science* 308: 264–67.
Linksvayer, T. A., and M. J. Wade. 2005. The evolutionary origin and elaboration of sociality in the aculeate Hymenoptera: Maternal effects, sib-social effects, and heterochrony. *Q. Rev. Biol.* 80: 317–36.

Oldroyd, B. P. 2002. The cape honeybee: An example of a social cancer. *Trends Ecol. Evol.* 17: 249–51.

Pankiw, T. 2004. Honey bee pheromones as information flow and collective decision-making. *Apidologie* 35: 217–26.

Pennisi, E. 2009. Agreeing to disagree. *Science* 323: 706–708.

Thorne, B. L., N. L. Breisch, and M. L. Muscedere. 2003. Evolution of eusociality and the soldier caste in termites: Influence of accelerated inheritance. *P. Natl. Acad. Sci. USA* 100: 12808–13.

Wheeler, W. 1911. The ant-colony as an organism. *J. Morphol.* 22: 307–25.

Wilson, E. O., and B. Hölldobler. 2005. Eusociality: Origin and consequence. *P. Natl. Acad. Sci. USA* 102: 13367–71.

Wilson, E. O. 2008. One giant leap: How insects achieved altruism and colonial life. *Bioscience* 58: 17–25.

Books and Reports

Bourke, A. F. G., and N. R. Franks. 1995. *Social evolution in ants.* Princeton: Princeton University Press.

Costa, J. T. 2006. *The other insect societies.* Cambridge, MA: Harvard University Press.

Darwin, C. 1859. *On the origin of species by means of natural selection, or the preservation of favoured races in the struggle for life.* London: John Murray.

Gadagkar, R. 2001. *The social biology of* Ropalidia marginata: *Toward understanding the evolution of eusociality.* Cambridge, MA: Harvard University Press.

Hölldobler, B., and E. O. Wilson. 1990. *The ants.* Cambridge, MA: Harvard University Press.

———. 2009. *The Superorganism: The beauty, elegance, and strangeness of insect societies.* New York: W. W. Norton.

Hunt, J. H. 2007. *The evolution of social wasps.* Oxford: Oxford University Press.

West-Eberhard, M. J. 2003. *Developmental plasticity and evolution.* Oxford: Oxford University Press.

Wilson, E.O. 1971. *The Insect Societies.* Cambridge, MA: Belknap Press.

———. 1994. *Naturalist.* Washington, DC: Island Press.

Interviews

E. O. Wilson. 2009. Interview by Roger Highfield. July.

Chapter 9: The Gift of the Gab

References

Boyd, R., and P. J. Richerson. 2009. Culture and the evolution of human cooperation. *Philos. T. Roy. Soc. B* 364: 3281–88.

Fitch, W. T. 2000. The evolution of speech: a comparative review. *Trends Cognit. Sci.* 4: 258–67.

Frith, U., and C. Frith. 2010. The social brain: Allowing humans to boldly go where no other species has been. *Philos. T. Roy. Soc. B* 365: 165–76.

Gold, E. M. 1967. Language identification in the limit." *Information and Control* 10: 447–74.

Jackendoff, R.1999. Possible stages in the evolution of the language capacity. *Trends Cognit. Sci.* 3: 272–79.

Komarova, N. L., and S. A. Levin. 2010. Eavesdropping and language dynamics. *J. Theor. Biol.* 264: 104–18.

Lieberman, E., J. B. Michel, J. Jackson, T. Tang, and M. A. Nowak. 2007. Quantifying the evolutionary dynamics of language. *Nature* 449: 713–16.

Matsen, E., and M. A. Nowak. 2004. Win-stay, lose-shift in language learning from peers. *P. Natl. Acad. Sci. USA* 101: 18053–57.

Michel, J. B., Yuan Kui Shen, Aviva Presser Aiden, Adrian Veres, Matthew K. Gray, The Google Books Team, Joseph P. Pickett, Dale Hoiberg, Dan Clancy, Peter Norvig, Jon Orwant, Steven Pinker, Martin A. Nowak, and Erez Lieberman Aiden. 2011. Quantitative analysis of culture using millions of digitized books. *Science* (advanced publication online, December 16, 2010).

Mitchener, G., and M. A. Nowak. 2003. Competitive exclusion and coexistence of universal grammars. *B. Math. Biol.* 65: 67–93.

Nowak, M. A., N. L. Komarova, and P. Niyogi. 2001. Evolution of universal grammar. *Science* 291: 114–18.

————. 2002. Computational and evolutionary aspects of language. *Nature* 417: 611–17.

Nowak, M. A., and D. C. Krakauer. 1999. The evolution of language. *P. Natl. Acad. Sci. USA* 96: 8028–33.

Nowak, M. A., D. C. Krakauer, and A. Dress. 1999. An error limit for the evolution of language. *P. Roy. Soc. Lond. B Bio.* 266: 2131–36.

Nowak, M. A., J. B. Plotkin, and V. A. A. Jansen. 2000. The evolution of syntactic communication. *Nature* 404: 495–98.

Valiant, L. G. 1984. A theory of learnable. *Communications of the ACM* 27: 436–45.

Books and Reports

Baker, M. C. 2001. *Atoms of language.* New York: Basic Books.

Bickerton, D. 2001. *Language and species.* Chicago: University of Chicago Press.

Chomsky, N. A. 1995. *The minimalist program.* Cambridge, MA: MIT Press.

Chomsky, N. M. 1957. *Syntactic structures.* The Hague/Paris: Mouton.

Dunbar, R. 1996. *Grooming, gossip, and the evolution of language.* Cambridge: Cambridge University Press.

Hauser, M. D. 1996. *The evolution of communication.* Cambridge, MA: Harvard University Press.

Hurford, J. R., M. Studdert-Kennedy, and C. Knight, eds. 1998. *Approaches to the evolution of language.* Cambridge: Cambridge University Press.

Jackendorf, R. 2002. *Foundations of language: Brain, meaning, grammar, evolution.* Oxford: Oxford University Press.

Lieberman, P. 1984. *The biology and evolution of language.* Cambridge, MA: Harvard University Press.

Niyogi, P. 2006. *The computational nature of language learning and evolution.* Cambridge, MA: MIT Press.

Pinker, S. 1994. *The language instinct.* London: Allen Lane.

Regis, E. 1987. *Who got Einstein's office?* London: Addison Wesley.

Smolensky, P., and G. Legendre. 2006. *The harmonic mind: From neural computation to optimality-theoretic grammar.* 2 vols. Cambridge, MA: MIT Press.

Tomasello, M. 1999. *The cultural origins of human cognition.* Cambridge, MA: Harvard University Press.

Vapnik, V. 1998. *Statistical learning theory.* New York: John Wiley.

Wexler, K., and P. Culicover. 1980. *Formal principles of language acquisition.* Cambridge, MA: MIT Press.

Interviews

Erez Lieberman. 2008. Interview by Roger Highfield. October.

Chapter 10: Public Goods

References

Bshary, R. 2002. Biting cleaner fish use altruism to deceive image—scoring client reef fish. *P. Roy. Soc. Lond. B Bio.* 269: 2087–93.

Dreber, A., and M. A. Nowak. 2008. Gambling for global goods. *P. Natl. Acad. Sci. USA* 105: 2261–62.

Fischbacher, U., and S. Gaechter. 2010. Social preferences, beliefs, and the dynamics of free riding in public good experiments. *Amer. Econ. Rev.* 100: 541–56.

Haley, K., and D. M. T. Fessler. 2005. Nobody's watching? Subtle cues affect generosity in an anonymous economic game. *Evol. Hum. Behav.* 26: 245–56.

Hardin, G. 1968. The tragedy of the commons. *Science* 162: 1243–48.

Hauert, C., S. De Monte, J. Hofbauer, and K. Sigmund. 2002. Volunteering as red queen mechanism for cooperation in public goods games. *Science* 296: 1129–32.

Hauert, C., J. Y. Wakano, and M. Doebeli. 2008. Ecological public goods games: Cooperation and bifurcation. *Theor. Pop. Biol.* 73: 257–63.

Killingback, T., M. Doebeli, and C. Hauert. 2010. Cooperation and defection in the tragedy of the commons. *Biol. Theor.* 5: 3–6.

Levin, S. 2010. Crossing scales, crossing disciplines: Collective motion and collective action in the Global Commons. *Philos. T. Roy. Soc. B* 365: 13–18.

Milinski, M., R. D. Sommerfeld, H-J Krambeck, F. A. Reed, and J. Marotzke. 2008. The collective-risk social dilemma and the prevention of simulated dangerous climate change. *P. Natl. Acad. Sci. USA* 105: 2291–94.

Myers, R. A., and B. Worm. 2003. Rapid worldwide depletion of predatory fish communities. *Nature* 423: 280–83.

Pfeiffer, T., and M. A. Nowak. 2006. Climate change: All in the game. *Nature* 441: 583–84.

Santos, F. C., M. D. Santos, and J. M. Pacheco. 2008. Social diversity promotes the emergence of cooperation in public goods games. *Nature* 454: 213–16.

Wakano, J. Y., M. A. Nowak, and C. Hauert. 2009. Spatial dynamics of ecological public goods. *P. Natl. Acad. Sci. USA* 106: 7910–14.

Books and reports

IPCC. 2008. *Fourth assessment report on climate change impacts, adaptation and vulnerability for researchers, students, policymakers.* Cambridge: Cambridge University Press.

Levin, S. A., ed. 2009 *Games, groups, and the global good.* Berlin: Springer.

Lovelock, J. 2009. *The vanishing face of Gaia: A final warning.* London: Allen Lane.

Ostrom, E. 2005. *Understanding institutional diversity.* Princeton: Princeton University Press.

Interviews

Tim Palmer. 2010. Interview by Roger Highfield. March.

Websites

The Garrett Hardin Society

www.garretthardinsociety.org

Background on Elinor Ostrom's Nobel Prize on the Nobel Prize website:

http://nobelprize.org/nobel_prizes/economics/laureates/2009/ecoadv09.pdf

Worldwide Fund for Nature, Living Planet Report

http://wwf.panda.org/about_our_earth/all_publications/living_planet_report

Chapter 11: Punish and Perish

References

Boyd, R., H. Gintis, S. Bowles, and P. J. Richerson. 2003. The evolution of altruistic punishment. *P. Natl. Acad. Sci. USA* 100: 3531–35.

Croson, R., and S. Gächter. 2010. The science of experimental economics. *J. Econ. Behav. Organ.* 73(1): 122–31.

de Quervain, D. J.-F., U. Fischbacher, V. Treyer, M. Schellhammer, U. Schnyder, A. Buck, and E. Fehr. 2004. The neural basis of altruistic punishment. *Science* 305: 1254–58.

Dreber, A., D. G. Rand, D. Fudenberg, and M. A. Nowak. 2008. Winners don't punish. *Nature* 452: 348–51.

Fehr, E., and U. Fischbacher. 2003. The nature of human altruism. *Nature* 425: 785–91.

Fehr, E., U. Fischbacher, and S. Gächter. 2002. Strong reciprocity, human cooperation, and the enforcement of social norms. *Hum. Nature* 13:1–25.

Fehr, E., and S. Gächter. 2000. Cooperation and punishment in public goods experiments. *Am. Econ. Rev.* 90: 980–94.

———. 2002. Altruistic punishment in humans. *Nature* 415: 137–40.

Gächter, S., B. Herrmann, and C. Thöni. Antisocial punishment across societies. *Science* 319: 1362–67.

Gächter, S., E. Renner, and M. Sefton. 2008. The long-run benefits of punishment. *Science* 322: 1510.

Hauert, C., A. Traulsen, H. Brandt, M. A. Nowak, and K. Sigmund. 2007. Via freedom to coercion: The emergence of costly punishment. *Science* 316: 1905–1907.

Helbing, D., A. Szolnoki, M. Perc, and G. Szabó. 2010. Evolutionary establishment of moral and double moral standards through spatial interactions. *PLoS Comput. Biol.* 6 (4): e1000758.

———. 2010. Punish, but not too hard: How costly punishment spreads in the spatial public goods game. *New J. Phys.* 12: 083005.

Henrich, J., R. Boyd, S. Bowles, C. Camerer, E. Fehr, H. Gintis, and R. McElreath. 2001. Cooperation, reciprocity and punishment in fifteen small-scale societies. *Am. Econ. Rev.* 91: 73–78.

Nowak, M. A., K. M. Page, and K. Sigmund. 2000. Fairness versus reason in the ultimatum game. *Science* 289: 1773–75.

Ohtsuki, H., Y. Iwasa, and M. A. Nowak. 2009. Indirect reciprocity provides only a narrow margin for efficiency for costly punishment. *Nature* 457: 79–82.

Ostrom, E., J. Walker, and R. Gardner. 1992. Covenants with and without a sword: Self-governance is possible. *Amer. Polit. Sci. Rev.* 86: 404–17.

Rand, D. G., A. Dreber, T. Ellingsen, D. Fudenberg, and M. A. Nowak. 2009. Positive interactions promote public cooperation. *Science* 325: 1272–75.

Rand, D. G., H. Ohtsuki, and M. A. Nowak. 2009. Direct reciprocity with costly punishment: Generous tit-for-tat prevails. *J. Theor. Biol.* 256: 45–57.

Sigmund, K., H. De Silva, A. Traulsen, and C. Hauert. 2010. Social learning promotes institutions for governing the commons. *Nature* 466: 861–63.

Sigmund, K., E. Fehr, and M. A. Nowak. 2002. The economics of fair play. *Sci. Am.* 286: 82–87.

Yamagishi, T. 1986. The provision of a sanctioning system as a public good. *J. Pers. Soc. Psychol.* 51: 110–16.

Books

Ostrom, E. 2005. *Understanding institutional diversity.* Princeton: Princeton University Press.

Chapter 12: How Many Friends Are Too Many?

References

Barabasi, A., and R. Albert. 1999. Emergence of scaling in random networks. *Science* 286: 509–12.

Erdős, P., and A. Rényi. 1960. On the evolution of random graphs. *Acta Math. Acad. Sci. H.* 5: 17–61.

Feld, S. 1991. Why your friends have more friends than you do. *Am. J. Sociol.* 96: 1464–77.

Fowler, J., and N. Christakis. 2008. Dynamic spread of happiness in a large social network: Longitudinal analysis over 20 years in the Framingham Heart Study. *Br. Med. J.* 337: a2338.

———. 2010. Cooperative behavior cascades in social networks. *P. Natl. Acad. Sci. USA* 107: 5334–38.

Fu, F., L. Wang, M. A. Nowak, and C. Hauert. 2009. Evolutionary dynamics on graphs: Efficient method for weak selection. *Phys. Rev. E* 79: 046707.

Grafen, A. 2007. An inclusive fitness analysis of altruism on a cyclical network. *J. Evol. Biol.* 20: 2278–83.

Gross, T., L. Rudolf, S. A. Levin, and U. Dieckmann. 2009. Generalized models reveal stabilizing factors in food webs. *Science* 325: 747–50.

Guimerà, R., B. Uzzi, J. Spiro, and L. A. Nunes Amaral. 2005. Team assembly mechanisms determine collaboration network structure and team performance. *Science* 308: 697–702.

Hill, A. L., D. G. Rand, M. A. Nowak, and N. A. Christakis. 2010. Emotions as infectious diseases in a large social network: The SISa model. *P. Roy. Soc. Lond. B Bio.*.

Ito, J., and K. Kaneko. 2002. Spontaneous structure formation in a network of chaotic units with variable connection strengths. *Phys. Rev. Lett.* 88: 028701–1.

Kühnert, C., D. Helbing, and G. B. West. 2006. Scaling laws in urban supply networks. *Physica A* 363: 96–103.

Lieberman, E., C. Hauert, and M.A. Nowak. 2005. Evolutionary dynamics on graphs. *Nature* 433: 312–16.

Newman, M. E. J. 2001. The structure of scientific collaboration networks. *P. Natl. Acad. Sci. USA* 98: 404–409.

Nowak, M. A., F. Michor, and Y. Iwasa. The linear process of somatic evolution. *P. Natl. Acad. Sci. USA* 100: 14966–69.

Ohtsuki, H., C. Hauert, E. Lieberman, and M. A. Nowak. 2006. A simple rule for the evolution of cooperation on graphs and social networks. *Nature* 441: 502–505.

Ohtsuki, H., M. A. Nowak. 2008. Evolutionary stability on graphs. *J. Theor. Biol.* 251: 698–707.

Ohtsuki, H., J. M. Pacheco, and M. A. Nowak. 2007. Evolutionary graph theory: Breaking the symmetry between interaction and replacement. *J. Theor. Biol.* 246: 681–94.

Pacheco, J. M., A. Traulsen, M. A. Nowak. 2006. Coevolution of strategy and structure in complex networks with dynamical linking. *Phys. Rev. Lett.* 97: 258103.

Skyrms, B., and R. Pemantle. 2000. A dynamic model of social network formation. *P. Natl. Acad. Sci. USA* 97: 9340–46.

Taylor, P. D., T. Day, and G. Wild. 2007. Evolution of cooperation in a finite homogeneous graph. *Nature* 447: 469–72.

Travers, J., and S. Milgram. 1969. An experimental study of the small world problem, *Sociometry* 32: 425–43.

Watts, D., and S. H. Strogatz. 1998. Collective dynamics of "small-world" networks. *Nature* 393: 440–42.

Books

Durett, R. 2006. *Random graph dynamics*. Cambridge: Cambridge University Press.

Articles

Wiseman, R. 2003. It really is a small world that we live in. *The Daily Telegraph,* June 4.

Chapter 13: Game, Set, and Match

References

Antal, T., H. Ohtsuki, J. Wakeley, P. D. Taylor, and M. A. Nowak. 2009. Evolution of cooperation by phenotypic similarity. *P. Natl. Acad. Sci. USA* 106: 8597–8600.

Nowak, M. A., C. E. Tarnita, and T. Antal. 2010. Evolutionary dynamics in structured populations. *Philos. T. Roy. Soc. B* 365: 19–30.

Tarnita, C. E., T. Antal, H. Ohtsuki, and M. A. Nowak. 2009. Evolutionary dynamics in set structured populations. *P. Natl. Acad. Sci. USA* 106: 8601–04.

Tarnita, C. E., H. Ohtsuki, T. Antal, F. Fu, and M. A. Nowak. 2009. Strategy selection in structured populations. *J. Theor. Biol.* 259: 570–81.

Taylor, P. D., and A. Grafen. 2010. Relatedness with different interaction configurations. *J. Theor. Biol.* 262, 391–97.

Books

Marx, G. 2009. *Groucho and me: The autobiography*. London: Virgin Books.

Interviews

Corina Tarnita. 2008. Interview by Roger Highfield. October.

Chapter 14: Crescendo of Cooperation

The translation of the opening quotation is as follows:

> The dear earth everywhere
> blooms in spring and grows green
> afresh! Everywhere and eternally,
> distant places have blue skies!
> Eternally . . . eternally . . .

References

Arrow, K., P. Dasgupta, L. Goulder, G. Daily, P. Ehrlich, G. Heal, S. Levin, K.-G. Mäler, S. Schneider, D. Starrett, and B. Walker. 2004. Are we consuming too much? *J. Economic Perspectives* 18 (3): 147–72.

Arrow, K., and S. A. Levin. 2009. Intergenerational resource transfers with random offspring. *P. Natl. Acad. Sci. USA* 106: 13702–706.

Beddington, J. 2010. Food security: Contributions from science to a new and greener revolution. *Philos. T. Roy. Soc. B* 365: 61–71.

Dasgupta, P. 2010. Nature's role in sustaining economic development. *Philos. T. Roy. Soc. B* 365: 5–11.

Fargione, J., J. Hill, D. Tilman, S. Polasky, and P. Hawthorne. 2008. Land clearing and the biofuel carbon debt. *Science* 319: 1235–38.

Levin, S.A. 1998. Ecosystems and the biosphere as complex adaptive systems. *Ecosystems* 1: 431–36.

———. 2006. Learning to live in a global commons: Socioeconomic challenges for a sustainable environment. *Ecol. Res.* 21 (3): 328–33.

May, R. M. 2010. Ecological science and tomorrow's world. *Philos. T. Roy. Soc. B* 365: 41–47.

May, R. M., S. A. Levin, and G. Sugihara. 2008. Ecology for bankers. *Nature* 451: 893–95.

Nowak, M. A. 2008. Generosity: A winner's advice. *Nature* 456: 579.

Nowak, M. A., and S. Roch. 2007. Upstream reciprocity and the evolution of gratitude. *P. Roy. Soc. Lond. B Bio.* 274: 605–609.

Pacala, S. W., and R. H. Socolow. 2004. Stabilization wedges: Solving the climate problem for the next 50 years with current technologies. *Science* 305: 968–72.

Rand, D.G., and M. A. Nowak. 2009. Name and shame: How reputation could save the earth. *New Scientist* 2734: 28–29.

Tilman, D. 2000. Causes, consequences and ethics of biodiversity. *Nature* 405: 208–211.

Tilman, D., J. Hill, and C. Lehman. 2006. Carbon-negative biofuels from low-input high-diversity grassland biomass. *Science* 314:1598–1600.

Books

Ridley, M. 2010. *The rational optimist: How prosperity evolves.* London: Fourth Estate.

Websites

Lugar, Richard G. 2005. *The Lugar survey on proliferation threats and responses.* http://lugar.senate.gov/nunnlugar/pdf/NPSurvey.pdf

INDEX

biology, xvi, 10, 12–15, 31, 46, 177, 226
  chaos in, 38
  dark side of, xi, xii, 157
  empirical, 104, 110
  mathematical, xvi–xvii, 4, 15–16, 38,
    40, 99, 147, 160–62, 165, 167,
    174, 256
  molecular, 4, 160
  population, 160
  systems, 228
  theoretical, 33, 65, 99, 131, 176, 262
  see also evolutionary biology;
    sociobiology
Biology of Moral Systems, The (Alexander), 60
biomimetic machines, 247
bipolar disorder, 26
birds, 25, 69, 151
  cries and songs of, 27, 180, 185
  see also specific birds
bison, 69, 202
black-and-white warbler, 185
Black Panther Party, 26–27
Blake, William, 39
Bliss, Mary, 99
blood, 95, 102, 272
  sharing of, 21–22, 45–46, 270
blood cancer, 149–50
blood cells, 151
Bloomberg, Michael, 82
Blumberg, Barry, 41
body, 12, 145
  cooperative reproduction of cells in,
    xiv–xv
  evolution of, 145
  organs of, xv, 139, 142, 151
Boltzmann, Ludwig, 9
Bonhoeffer, Sebastian, 274
Bonner, John, 140–41
Bowles, Sam, 90
Boy Scouts, 61
Bozic, Ivana, 150
brain cancer, 189
brains, 58, 139, 171
  cognitive capacity of, 25
  decision making by, 101
  development of, 56, 171, 172
  evolution of, xvii, 172

language and, 171, 172, 177, 185–86,
    197
  neurological networks of, 239, 242,
    275–76
  reward centers of, 222
  size of, xii, xiv, xvii, 177, 185–86, 197
  structure of, 275–76
brain scans, 222
Brand, Stewart, 70
Brandt, Hannelore, 66
Brave New World (Huxley), 96
breast cancer, 279
Bronson, Charles, 222
Brown, Robert, 115–16, 120
Bruckner, Anton, 279
Buddhism, 56
Buridan's ass paradox, 32
business, 231, 247
  cheating and collusion in, 8, 14
  competition vs. cooperation in, xii, 8
  contracts in, 25
  dream teams in, 251
  rise and fall of, 37
Butterfield, William, 174

Calabi-Yau manifolds, 256
calculus, 78
Cambridge University, 15, 23, 70, 72
cancer, 246
  age and, 144
  breakdown of cellular cooperation in,
    141–49, 151–52
  cellular development in, 142–50, 245
  deaths from, 141, 150, 279
  lifestyle choices and, 144–45
  metastasis of, 143
  treatment and prevention of, xvii, 143,
    145, 146, 150–52, 247
  see also specific cancers
canoe design, 91
Cape honeybee, 157
carbohydrates, 116
carbon dioxide (CO_2), 199, 206, 210,
    214, 218
carcinogens, 144
carcinoma, 143
cats, 139, 205

ABOUT THE AUTHORS

MARTIN A. NOWAK is Professor of Mathematics and Biology at Harvard University and Director of the Program for Evolutionary Dynamics. Nowak studied biochemistry and mathematics at the University of Vienna, where he received his PhD summa cum laude in 1989 studying with Karl Sigmund and Peter Schuster. Afterward, he went to Oxford to work with Robert May (now Lord May of Oxford). Nowak became Professor of Mathematical Biology at the University of Oxford at the age of thirty-two. In 1998 he moved to Princeton to establish the first Program in Theoretical Biology at the Institute for Advanced Study. In July 2003, Nowak was recruited by Harvard University as a full professor.

Nowak has been a leading light in the effort to apply mathematics to biology. In cooperation with many scientists over the years he has discovered some of the fundamental principles according to which life unfolds. Nowak has developed concepts and research areas such as prelife, virus dynamics, evolutionary genetics of cancer, spatial games, stochastic games, indirect reciprocity, evolutionary graph theory, and the mathematics of language evolution. He is the author of more than forty papers in *Nature, Science,* and *Scientific American.* In total, he has published more than three hundred papers and also three books. Nowak is well known among his peers for the deceptive simplicity of his models.

ROGER HIGHFIELD, DPhil, studied for his doctorate at Oxford University and the Institut Laue-Langevin, Grenoble. He is Editor of *New Scientist* magazine, which is now the world's biggest-selling weekly science and technology magazine. Prior to joining *New Scientist,* he was the Science Editor of *The Daily Telegraph,* where he worked for more than twenty years and still contributes a column. Highfield has been a regular contributor to BBC radio, sat on a range of advisory committees, and won many awards.

He has written/coauthored six popular science books, two of which have been bestsellers, including *After Dolly, The Science of Harry Potter, The Physics of Christmas, The Private Lives of Albert Einstein, The Arrow of Time,* and *Frontiers of Complexity*—all of which have been translated into foreign editions. His most recent work was as the outside editor on genomic researcher J. Craig Venter's autobiography, *A Life Decoded,* published in November 2007 (Viking, US; Allen Lane, UK).